A WORD FROM OUR VIEWERS

A WORD FROM OUR VIEWERS: REFLECTIONS FROM EARLY TELEVISION AUDIENCES

Ray Barfield

Westport, Connecticut
London

Library of Congress Cataloging-in-Publication Data

Barfield, Ray E.
    A word from our viewers : reflections from early television audiences / Ray Barfield.
        p.   cm.
    Includes bibliographical references and index.
    ISBN 978–0–275–99870–7 (alk. paper)
    1. Television broadcasting–United States–History.   2. Television viewers–United
States–Attitudes.   I. Title.
    PN1992.3.U5B35   2008
    302.23′450973–dc22   2007029032

British Library Cataloguing in Publication Data is available.

Library of Congress Catalog Card Number: 2007029032
ISBN-13: 978–0–275–99870–7

First published in 2008

Praeger Publishers, 88 Post Road West, Westport, CT 06881
An imprint of Greenwood Publishing Group, Inc.
http://www.praeger.com

Printed in the United States of America

The paper used in this book complies with the
Permanent Paper Standard issued by the National
Information Standards Organization (Z39.48–1984).

10 9 8 7 6 5 4 3 2 1

To Beth, Kelley, and Emily

Contents

Contents

Preface

Television is the device that spits out programs sucked in by a cable or satellite connection and plays them in real time or delays them through further connection to a VHS, TiVo, or other recording attachment. Right? It is the electronic instrument that, with a flick of the remote, will give you *Oprah* or the weekly episode of your prime time favorite or the network evening news or anytime news from CNN or Fox. Each time you replace the TV, the screen grows larger, the case becomes thinner, and you are confronted with more and more terms and initials to decipher in the electronics marketplace: Flat-screen, HDTV-ready, plasma, and many more to come. Isn't that, or something like it, your recent experience of television?

The first TV in your life was also perhaps the first in the neighborhood, and it arrived with plenty of visitors, wanted or unwanted. If it was a 1940s black and white or a 1960s color set, it came with its own attendants, who unpacked it, tuned it to a test pattern or to their own perceptions of the best picture, and left instructions and admonitions for its continued use. Then you were left on your own to deal with small knobs marked VERT and HORIZ and a larger one listing channel numbers 2 through 13, to be manually selected. The set was tethered to an outdoor antenna, with a rotor for the carriage trade, or it made do with telescoped rabbit ears and (for UHF reception on later models) a loop that always got in the way. Or, even before that, your first glimpse of TV occurred at the window of an appliance store or a radio-TV shop where a set had been left to play after closing time. In your home it became a living comic page or a news broadside, a place of grieving at the time of the Kennedy and King assassinations, a ready companion from early morning sign-on to midnight sign-off. Are these your memories too?

Television *is* . . . what?

Television *was* . . . what?

Cecelia Tichi notes that one could sense the division of two television generations by watching the crowds at the 1989 Smithsonian Institution exhibition "American Television: From the Fair to the Family, 1939–1989": "To stand at a railing before a display of early TV receivers was to hear older people recite in considerable detail—almost compulsively—their recollections of the first television broadcast they saw, the first TV set to be brought into their habitat. At length they would describe the cabinetry, their first programs, the recollected remarks of family members or friends in the immediate vicinity." Children and grandchildren accompanying these oral historians were "distinctly unimpressed, enduring the TV stories in apparent boredom." The elders in this tableau had "consciously experienced the introduction—even incursion or intrusion—of television into their lives and habitats and, therefore, understood it to be discontinuous...," while the younger attendees "never knew life without the small screen and have experienced television as integral and natural.... Built in—and thus unremarkable."[1] Bill Bryson, in a memoir of his 1950s youth, says, "It is almost not possible now to appreciate just how welcome TV was."[2] Indeed, a faint note of disbelief can be detected in Ryan Van Cleave's voice when he remarks, "My parents told me they once drove forty miles across Iowa to see...*Howdy Doody*." Of course the purpose was not specifically to see Bob Smith's puppets but to see *television*, and Howdy happened to be the only representative available at the time.

Replete with images of stars falling over program titles, slapstick blasts of seltzer water, and full-orchestra fanfares, early television sought to persuade its viewers that the small black and white screen was in fact a huge spangled canvas where they could witness not merely any entertainment but staged extravaganzas worthy of being called *Your Show of Shows* or, as Ed Sullivan said in near self-parody each week, "Our real-ly BIG shew." The mask surrounding the picture tube echoed the proscenium frame of the stage, and variety shows as well as dramas were trumpeted as "coming to you" from the site of one entertainment venue or another: a theater in Hollywood or a Broadway playhouse. Early watchers saw the medium as a wonderment, a bit of "magic," a novelty, a heaven of renewable delights. But the intense hues of the dawn (even when viewed on a ten-inch black and white cathode tube) must fade into the common light of day. Over several decades of reporting assassinations, wars and war protests, and corruptions in business and government, television has lost that sense of theatrical framing. New large-screen monitors are bordered by modest black strips rather than sculpted frames, and the increasing portability of video cameras has meant that contemporary viewers are more often thrust onto the steamy, dangerous street than into the theater balcony. Simulated "magic" has been replaced by sometimes icky and often tweaked "reality," where home viewers ask themselves or each other, "For $10,000, would I (or you) eat the bug that guy is chewing on?"

"In its so-called Golden Age," Alessandra Stanley observes, "television served as a kind of social training film; sitcoms like *Father Knows Best* or *Leave It to Beaver* taught Americans emerging from World War II how to be middle-class." Most of today's prime-time "reality" programming, Stanley continues, "is instructional in an atavistic way: The focus is less and less on the middle class and more on the wealthiest 1 percent at the top and the masses of striving people at the bottom."[3] Thus *American Idol* echoes the desperate-to-get-into-show-business plots of Depression-era movies, while *The Apprentice* allows viewers in the early 2000s to look up to Trump Tower (and to eavesdrop on the clawing and conniving efforts to gain executive positions therein) as their grandparents craned their necks to see the tops of the Chrysler or the Empire State Building from street level.

Columnist Leonard Pitts contemplates the case of Ryan Sutter, a firefighter and paramedic who was married to Trista Rehn on ABC-TV's new-style reality show *The Bachelorette* and since then has been beset by autograph seekers—interrupted (in Pitts's words) by "Some teenybopper in full gush"—even while he is treating an accident victim at a fire scene or on a roadside. Speaking of "Our Pavlovian relationship with the television camera" and thinking of Sutter's rueful loss of privacy, Pitts wonders if today's television "reality"—in which contestants subject themselves to a variety of humiliations and where even sign-waving visitors to the *Today* show seem to seek the existential validation of being glimpsed on TV—is better than the entertainments offered to pioneering viewers of the 1940s, 1950s, and 1960s:

> We invent television and television invents us, tells us what to want and who to be. It's an endless cycle that has spawned a pop culture brimming with shallowness, emptiness and a stupidity so profound as to be unprecedented. I say this as someone who's seen every episode of *Gilligan's Island*.
>
> The difference between that era and this one is that then, you usually became famous for something you did—acting, for instance. But these days, you're just as likely to become famous for something you "allow," for your willingness to let a TV camera poke, with proctoscopic invasiveness, into the most intimate regions of your life.[4]

Perhaps Pitts has forgotten *Queen for a Day* (not NBC-TV's proudest achievement in the 1950s), in which ever-cheerful Jack Bailey probed the financial or emotional needs of three contestants each day, or the later quiz show scandals, where shame proved to be the price of brief fame. His observations nonetheless usefully mark a general distinction between pioneering viewers' TV fare and today's video parade.

Each television generation, then, will define the medium in its own terms. The young and the indifferent are free to have it their shrugging way, but their elders' memories—subject, true, to gaps of recollection or misremembering—tell us of the beginnings of a major and (like it or not) pervasive medium and of the early audiences that helped to direct its evolution by being counted in the Nielsen ratings, by buying sponsors' products, or in strongly expressing their programming likes and dislikes to network and local station officials, to editors of newspaper TV sections, to the Federal Communications Commission, or to each other. They first knew it in a simpler time, and they found it a source of great delight, a companion for a lonely childhood or a span of old age, a formula-ridden curse, an electronic firefly.

The largest part of this book centers on the "TV tales" of that earlier generation of viewers who cannot quite give up the habit of referring to "the TV" as a "TV *set*," and the early chapters constitute a forum in which they convey their experiences of TV receivers and paraphernalia, viewing companions, programs and performers, and other aspects of the TV life. Later chapters offer a larger context for those recollections through reference to artistic works, the screeds and measured ruminations of media observers, and other long-view material. Primarily an oral history account, this study is not meant to be a directory of TV programs or a season-to-season chronicle of television programming, and paragraphs of jargon-filled abstractions and sociographed numbers have been left for others to write.

Conjuring up a forum of veteran viewers requires a great deal of generous help. First to be thanked are the watchers themselves, who communicated by letter, e-mail, telephone interview, taped session, and face-to-face conversation. They are credited where they are quoted except in a very few places where they have opted for anonymity. Beyond contributing their own recollections, some, including Corinne Holt Sawyer, Teresa Williams, Bernadette Longo, Linda Regnier, and Daniel Galvin, have prompted friends and relatives to add their stories.

Still others have helped to spread the word about this project. Jeanne Brooks generously invited her *Greenville News* column readers to contribute material, while Tim Anderson and Sherri Butler were similarly helpful in *The Herald-Leader*. Clemson University campus publications editors Robin Denny, Beth Jarrard, and Liz Newall efficiently posted announcements to staff and alumni, who were predictably responsive in offering a wide variety of viewing tales from many settings. Charlotte Holt conducted an interview for the ETV Radio show *Your Day*, and veteran broadcaster Wally Mullinax arranged a meeting with the WFBC-TV Old Timers. Jim Snyder made room for a project announcement in the *NARA News*, Barbara Watkins placed a notice in the *SPERDVAC Radiogram*, and Owens Pomeroy usefully tapped his grapevine of broadcasting enthusiasts.

Summer research assistants Jon DeWorken, Amy Gibb, Allison Kellar, Shawn Pressley, Allen Swords, and David Yurko provided diligent and resourceful help in tracking down obscure details and sources, while Clemson English Department administrators and colleagues Frank Day, Martin Jacobi, Bill Koon, Mark Charney, Alma Bennett, Deb Morton, and Sean Williams gave their support in many ways, as did Pearl Parker, Grace Ammons, and Christopher Greggs. Thanks are also due to Dean Jan Shach of the College of Architecture, Arts, and Humanities for facilitating a sabbatical semester.

Frank Chorba, Jim Freeman, and M. Thomas Inge have given their sustained professional encouragement for many years now, and Lloyd Bryant has generously shared his observations. Daniel Harmon at Praeger Publishers has been most helpful.

My deep thanks go to all.

Notes

1. Cecelia Tichi, *Electronic Hearth: Creating an American Television Culture* (New York: Oxford University Press, 1991), 8.

2. Bill Bryson, *The Life and Times of the Thunderbolt Kid: A Memoir* (New York: Broadway Books, 2006), 74.

3. Alessandra Stanley, "Class Issues: TV's Busby Berkeley Moment," *New York Times* (January 30, 2005): sec. D.

4. Leonard Pitts, "Humans Can't Bear Much Reality," *Herald-Leader* (Fitzgerald, GA) (June 30, 2004): sec. B.

First Watchers: An Introduction

In the East, the Sheppards watched by night. After an outlay of $169.50 in 1948 dollars, they gazed each evening at the partially masked seven-inch round screen of their Hallicrafters model T54 and played with the front panel punch-buttons for channels 1 (yes, 1) through 13, though there were only two active channels to see. Their bridge-foursome friends, the Evanses, often watched with them, arriving just at that late afternoon half hour when the test pattern yielded to the first program and going home (next door) soon after the sign-off four hours later. Within a few months the uninvited neighbor Sol Sorensen—a man who could replace a ballast before some people could decide how to spell it—finished assembling his first Heathkit TV monitor in his workshop around the corner, and the Sorensens watched from the new early movie show until late evening. Two years later the postwar subdivision's streets snaked past ninety homes where nearly half of the families watched by daytime and evening hours, and exercise walkers could see the blue-gray glow in the rectangular living room windows of house after house down the street. Evening strollers grew fewer as more and more residents sank into comfortable viewing positions immediately after dinner—or *for* dinner served on flimsy aluminum trays. Not many years after that, the subdivision itself crept over the crest of the hill and down the slope, and most of the 900 then residing there had the options of rising with the television day's first folksy homemaking program at 6:30 A.M. and of snoozing in the La-Z-Boy or on the sofa until after the military academy band's filmed rendition of the National Anthem had put a period to the solemn words of the station's 1:00 A.M. sign-off. TV owners in their recliners woke suddenly in the middle of the night, with only a rushing roar greeting them from the speaker. The TV age had certifiably begun, and postwar America was being tuned to it.

Two decades later the Evanses became pioneer time-shifters with their new Betamax video recorder and its initially pricey blank tapes (nearly $20 each), and the Sheppards had decommissioned the antenna and signed on with the new cable outfit at the monthly rate of $9.98, safely regulated, apparently for all time. With the connivance of the babysitter, the Sheppard grandchildren would overleap the access code and discover some revelatory TV on the Playboy channel. The Sorensens, having graduated to a Heathkit color monitor, would come to realize that they were watching more television and enjoying it less. Still later Hubert and Kathy Evans, resettled in Ft. Lauderdale, yielded control of their satellite-fed, wall-mounted plasma screen monitor-computer-TiVo remote to the more dexterous hands among the three generations of family passing into and out of their retirement cottage. They found no need to keep up the *TV Guide* subscription or to save the Sunday newspaper's weekly small-type program listings; the on-screen satellite guide offered a continuous parade of TV viewing options, and the remote control acted as an electronic index finger to open up seemingly endless choices. TV wasn't just TV anymore, except on the nostalgia channels, where the viewer might be prompted to dream of *Playhouse 90* and *Lost in Space* and Julie's sixth birthday, when she had made faces to the camera from the local kiddie show's Peanut Gallery, back in television's first—some say its Golden—age.

—— I ——

How They Watched

— 1 —

Anticipations and First Sightings

About 1951 young Zelime Lentz heard her school friends and neighbors talking about "televisions," and she asked her father, "What are they?" He responded, she recalls, "that television was a radio with a movie [added] and that we would never get one because they were just for rich people."[1] To the contrary, "Lemie" Lentz belonged to a generation that would watch the fully tubed black and white console receiver dominating a corner of the living room, then the transistorized portable purchased at the discount store on a whim, later the small-screen kitchen counter AM/FM/TV set whose little speaker competed with sounds of chopping and frying. In time the new color set, circa 1962, would replace "the old unreliable," and the typical household's second receiver would stand near the foot of the bed so that late evening viewers could catch Johnny Carson's recurring *Tonight* show joke about being watched across the country by people whose view was framed by pairs of naked feet.

The often delayed promise of TV's "imminent" arrival teased most would-be viewers in the 1930s. Some radio sets of the period included a "Television" option among their tuning punch buttons, and many a child gave that button a push, hoping that a picture would magically appear somewhere. Splashy cover paintings for pulp science fiction and hobbyist magazines, especially those edited and published by Hugo Gernsback, imagined TV pictures projected into the air, dancing atop strutting cones as startled onlookers stood in openmouthed admiration. In Schenectady, New York, a small audience saw the first-ever televised play, *The Queen's Messenger*, in 1928. In Chicago two years later, ads for WMAQ's experimental station W9XAP promised "TELEVISION TONIGHT and *every* night," and in early 1931 a *Universal Newspaper Newsreel* showed *Amos 'n' Andy* announcer Bill Hay reading a script before that station's TV camera in the *Daily News* building. For the few who had access to monitors, there was television; for the many who did not, there were rumors of television.

Newspaper readers in the novelty-loving Depression era saw occasional accounts of TV experiments in laboratories and at transmitter sites, and financial pages recorded patent applications for video circuitry by Philo Farnsworth and the bitterly competing RCA labs, but radio remained the chief broadcasting supplier of "pictures" from the larger world. "I can remember my grandfather telling me as we listened to the radio that someday we would be able to see people instead of just listening to them. It all seemed strange to me at the time," Marion F. Dilger says, adding that in the meantime she and her family were content with hearing *The Shadow, Fibber McGee and Molly*, and *Baby Snooks*, the last of these supplying the "Snooky" nickname "that my dad called me for many years to come." Indeed, TV's arrival would be largely deferred between 1941 and 1945, when U.S. electronics interests were redirected toward the World War II victory effort. Nonetheless, Roger Rollin recalls, "Even back in the '30s we kids had heard of television. I seem to remember a 'Nancy and Sluggo' comic strip involving television even though it didn't really exist [in everyday public experience] then. (This strip was always curiously devoid of context: a-historical, even a-geographical.) And I seem to remember something like television in 'Flash Gordon' Sunday sci-fi (then labeled 'futuristic') comics and in *Buck Rogers* Saturday afternoon movie serials. The 1939 World's Fair in New York was a very big deal at the time, and weren't there demonstrations of rudimentary TV there?"

In fact, an early television system was shown at the Chicago World's Fair in 1933 and 1934, but it was almost lost in the general variety of that Century of Progress Exposition. Nonetheless, Walter Cronkite, later to anchor *The CBS Evening News* and to be called "the most trusted man in America," remembers his first glimpse of the medium. In 1933 he set out "with a group of us who saved our money and went up from Houston, Texas, to Chicago, and they had an exhibit of television there. I was one of those who volunteered to have my picture taken, and it was shown on the other side of the room with great amazement of everybody present."[2]

In 1933, members of the Cox family in Mt. Olive, North Carolina, also encountered that television display. A sizable family group, including H. Morris Cox's parents, uncle and aunt, cousins, and a neighbor or two who bargained their way into the trip, made their way to Chicago in a 1932 Ford touring car and a large Studebaker truck (ordinarily used for carrying up to eight mules), and in three days of travel they reached the Fair site. The parents soon heard that the exotic ecdysiast Sally Rand was performing her celebrated act at the Fair, and, as family historian John N. Cox recounts, they "specifically prohibited the boys from even walking past it." Morris Cox, the oldest of the cousins, "could have passed for 21" and used his deepest voice to ask the ticket seller for three tickets, but the supervising barker forced a confession that the boys were younger than the posted age limit, "reproved the ticket seller," and sent the Cox boys on their way to sample other sights of the Fair, including a German diesel locomotive and a $25,000 Duesenberg. John N. Cox notes that they "repeatedly went to see one of the first television sets ever displayed in public. At each television demonstration, a person from the audience was invited on stage to stand in front of a camera and appear on television. At one show, Bob was selected and became the first person in the Cox family to appear on TV." H. Morris Cox recalls that participants were warned that their television images would be somewhat fuzzy and flickery, as indeed they proved to be.

In the next World's Fair at the end of the decade, television was presented not only as a technological marvel but also as an emerging entertainment medium, with NBC supplying limited programming for about 200 home sets in New York City as well as for visitors to "The World of Tomorrow" at Flushing Meadows in Queens. When Eloise Beckett's parents took her to the 1939 New York World's Fair, she was most impressed by two exhibits: the General Motors Futurama building's "Highways and Horizons" model of a projected 1960s "super highway" and, at the RCA Exhibit Building, the first public demonstrations of programmed television. In a courtyard behind the vacuum tube-shaped RCA building Eloise and the other members of the family were able to see tiny pictures of themselves on three- or four-inch black and white screens. Like other visitors, they "spent a lot of time just making faces into the camera and whispering to each other." Upon leaving, each guest received a wallet-sized card reading, "This is to Certify that——has been TELEVISED at the RCA EXHIBIT BUILDING, NEW YORK WORLD'S FAIR,——1939," authenticated, so to speak, by the printed signature of the exhibit manager. The "Buck Rogers fantasy dream of future roads," Beckett laughingly observes, eventually became the tangle of "spaghetti junctions" in cities linked by the Eisenhower Interstate Highway System, while television proved to be a different path to the future.

Having the right social connections in the late 1930s meant getting an early private glimpse of the new medium and, on a good day, seeing a TV "first." Dick Eyman was a friend of John Runyon, a CBS vice-president's son, and was invited to the Runyons' home in Darien, Connecticut, on May 17, 1939, to witness the first athletic contest ever televised, a college baseball game pitting Columbia University (the home team at Baker Field) against Princeton. Bill Stern was the announcer for W2XBS's 4:00–6:15 P.M. telecast.[3] This was "just about the time that TV was publicly introduced at the New York World's Fair," Dick Eyman recalls, and the private viewing took place through "one of the early home TVs where the picture was reflected onto a screen underneath the lift-top of the console."

Dave Milling had a more sustained view of the medium in those days when sets were closely rationed:

> We lived in New Jersey until 1943, and my dad, "Shine" Milling, was with RCA in Camden. After RCA first displayed TV to the public at the 1939 World's Fair, a few sets were available for evaluation by some RCA employees. We were fortunate to have one of these early sets. It, like all sets at the time, was a large cabinet with a lid that lifted, exposing a mirrored surface. The TV tube was mounted vertically in the cabinet, so you viewed the image on the mirrored lid. It was about a 7-inch picture.
>
> There was only one commercial channel—WPTZ, Philadelphia. There were programs three times a week. On Monday and Wednesday nights, there was a movie, but on Friday nights there was a movie *and* a continuing serial.

Sets in the Philadelphia area could probably be counted in the hundreds or less, and every household with a set received a postcard weekly with the programs listed. I have a very distinct recollection of my dad holding up the radio program listings from the *Philadelphia Inquirer* and telling me and my sisters that someday the *Inquirer* would list TV programs and we would no longer receive a weekly postcard.

In 1943 my dad was transferred to Indiana, and there was no more TV for us until after the War. Soon after the War ended, we were transferred back to New Jersey, and we actually purchased a TV set. Programs *were* listed in the paper, and everyone's favorite was Milton Berle's show, *The Texaco Star Theatre.*

Milling's subsequent memories include "checking vacuum tubes at the 7-11 store," seeing the first color sets, and watching early Japanese-manufactured receivers begin to displace domestically produced ones: "It used to be RCA, Zenith, Emerson, Philco, Curtis Mathes, Muntz, Magnavox, etc., etc."

G. B. "Bill" Seaborn, attending a U.S. Navy radio technicians' school in Chicago in the summer of 1943, noticed that in another part of the State-Lake building engineers were experimenting in ways to improve TV transmission quality. Pleased that he was "able to see firsthand some of the work that was going on," he observed that "in one room a small stage was set up with lots of lights and one or two TV cameras. In an adjoining room there was a TV set, about a 5-inch as I remember. From time to time performers, mostly dancers, would go through a routine on the stage while technicians would view the results in the room next door. It was fascinating to be able to see the live performance and the TV reproduction simultaneously."

Ethel L. Morris found the programming sparse and the reception problematic in her first wartime encounter with the medium:

I lived on Long Island, in the shadow of New York City, and during the summer of 1943 worked at Sperry Gyroscope Co. in Manhasset. In addition to the vital gyroscope development, something else was new at Sperry. A few of the young engineers were building their own television receivers for the once a week broadcasts sent from the city.

I was thrilled with the invitation to a TV party to watch a wrestling match sponsored by Gillette. But due to inconsistencies and distance between sending and receiving stations, the pictures were often distorted and looked as though they were seen through water. Although most of us were not wrestling fans, the novelty of this new concept was exciting!

Then I finished school, married, and moved West to the Rocky Mountain Empire, where no one we knew had television. By the time I returned to New York in 1949, my father had bought an expensive TV set and each evening watched the national news broadcasts followed by a delightful program featuring Fran Allison and her puppet friends Kukla and Ollie.

A self-described "country boy from Bluffton, South Carolina" who grew up "in a very isolated area without even electricity or telephones," John C. Pinckney first encountered television during a tour of Radio City in New York in 1945, just before he entered college: "This included demonstrations of live TV where the guests could see the performers in the studio and at the same time view the scene on a nearby monitor." Three years later, before many citizens had glimpsed television at all, he had two further memorable encounters with the medium:

> In 1948 while attending ROTC summer camp at Ft. Mead, Maryland, I witnessed a live broadcast of a heavyweight fight between Joe Louis and, as I recall, Jersey Joe Walcott. The screen was 6 or 8 inches wide and so snowy that one had to be at least 10 feet away to make heads or tails of the picture. It was the only TV set in the area, and there were literally dozens of cadets crammed together to see what I am sure was their first exposure to the new medium.

In November, he and the other members of the Clemson Senior Platoon performed a drill during half-time at a professional football game in Yankee Stadium. Unaware that their routine had been telecast, they were surprised later when, as they were "rubber-necking in Manhattan, we were approached by several natives who had seen us! Talk about country-come-to-town!" Even today he sees "TV and all the other modern electronic gadgets we have nowadays" as still "somewhat a wonder and not taken for granted."

When the first commercial television stations began to dot the country, transmitting programs from the late afternoon until late evening, customers entering drinking establishments found that most eyes were turned to the bracket-mounted set at the end of the bar. (In a joke from that period Henny Youngman sadly asserted that his brother-in-law had lost his bartender's job because he was inept at keeping the TV set in repair.)[4] World War II veteran Jim Hattaway notes, "My first TV was seen in a bar in 'Frisco after 20 months on Guam," while John F. Clark recalls, "In the '40s my boss took me with him to a meeting at the company head office in Pittsburgh. One evening we went into the Men's Bar of the William Penn Hotel, and there we saw our first television show. It was *Howdy Doody*. We didn't stay to the end but agreed that there was no future for television." What John Clark and his boss had witnessed was, in fact, the beginning of network programmers' strategy to shift postwar viewing from saloons and hotel bars to home receivers. Like Bob Smith's *Howdy Doody Time*, Burr Tillstrom's Kuklapolitan Players, newly teamed with Fran Allison, had a large role in that change. In 1947, according to Susan Gibberman, the Kuklapolitans were chosen as "the perfect 'family fare,'" and the planned thirteen-week run stretched to ten years of entertainment that gave many a household the impetus (or the excuse) to invest in a television set.[5]

Corinne Holt Sawyer, who would later host a TV homemaker show in North Carolina, recalls another early mismatch of audience and programming:

> I saw my first TV show in Miami in 1948. There was a fancy cocktail party in a private home, and the host showed us that he had acquired a TV set, which he turned on for the guests to marvel at. The set was smallish—perhaps 30 inches high by 18 inches wide (and very deep), and it had a small round screen set high in its grey metallic

case; the screen was perhaps 9 inches in diameter, and the picture (black and white, of course) was grainy and filled with electronic "snow." The set looked more like today's industrial monitors than like today's domestic TV sets, and our host had set it up on a low end table, since he had no other stand for it.

Miami was not then linked to New York by coaxial cable, and live shows were all local. The alternatives were to show film or kinescope, and the presentation our host tuned in (he said he could get only one channel) was of an old episode of *The Lone Ranger.* The guests gathered around in their fancy clothing, still clutching drinks and plates of food. I remember that I was wearing a dress of ice-blue satin that creased badly when I sat for very long, and I had planned to stand and circulate (as one does at cocktail parties) the whole time and thus preserve that 'band box' look. But I, like all the others, ended sitting down on the floor, regardless of creased clothing and spilled food and drink, crowded uncomfortably close together so we could see that tiny screen—and we watched that entire episode of *The Lone Ranger* from first to last, absolutely mesmerized, until the host, desperate to get his stalled party going again, turned the set off.

Perhaps it was that experience, with everyone (including myself) paying such close attention to the screen, regardless of content, that convinced me that while TV might be good evening entertainment, it would never catch on for daytime viewers. How could business people watch a TV show during the day? How could a housewife get her dishes washed and her clothes ironed and watch the screen at the same time? Radio would remain the dominant medium during daytime hours, I believed, and so I told students in my "History of Broadcasting" class at the University. Talk about your "clouded crystal ball"!

Big city residents naturally had an advantage in first TV sightings. When musician John Butler left his home in Chicago to begin his freshman year at West Texas State in 1951, he "was regarded as being very sophisticated. I had *seen television!*" he recalls with typical bemusement.

For the general postwar population, isolated viewings of television gradually yielded to visit-and-watch rituals among relatives and neighbors. John B. Ashe says,

My earliest recollection was in the mid-'40s, watching baseball games with my dad at our home in Rock Hill, South Carolina. After marriage in December 1948, my wife and I had an apartment across the street from a couple our age, good friends who lived with the wife's uncle and aunt. I recall how on Saturday nights the six of us, and usually a few other neighbors, would sit around the large screen (17-inch) TV in a semi-circle, watch *The Lucky Strike Hit Parade*, and marvel over the animated cigarette packs [dancing in the commercials]. And of course we had to watch Milton Berle, Jackie Gleason, Imogene Coca, Sid Caesar, and many others.

Those who lacked television-set-owning relatives soon found other sites. In Baltimore, Owens Pomeroy says, "The neighborhood fire department houses were the center of TV entertainment for the surrounding area, even before there were sets in the homes. Many people, out for a summer stroll, often took a detour to the firehouse to catch the latest show on TV." In South Carolina, the newly opened Clemson House hotel advertised itself as "swank," and one feature of that putative swankiness (along with the now long-departed French chef and the dances featuring leading bands of that day) was to have a television set in the penthouse, where, as Robert F. Mixon remembers, "Townspeople (young and old) used to go on Saturday nights to watch *The George Gobel Show* and *Wide World of Wrestling*." Jerry van de Erve, a class of 1930 college graduate, "well remember[s] when TV came in" and notes that in Seattle in the late 1940s his daughter, "born in 1944, every afternoon about 5:00 would join at least a dozen of her friends at a house two blocks away to watch a children's program on a 5-inch TV."

In most localities, the department or appliance store was the place of resort for the TV-disenfranchised. One viewer "vividly" recalls the 1951 Tuesday evening ritual in New Jersey: "My mother and I would stand outside the local appliance store to watch Uncle Miltie. The sets were all turned on, and the sound was piped through a loudspeaker. There was always a good crowd no matter how cold the weather." In Sioux City, Iowa, Marilyn Chadwick found her viewing spot in what seemed to be a scaled-down drive-in movie setting: "A local electronic and appliance business by the name of Gibson's had a large plate glass window facing the parking area in the front of their building. As a marketing strategy, during the evening hours when the store was closed, a TV was left on for public viewing. Dad, Mom, my sister (Diane), the dog, and I would go for an ice cream cone and then wheel into Gibson's and watch this innovative device from the comfort of our own car."

Gambles Home and Hardware on East Sixth Street in Des Moines made quite a show of its "window movies" on Saturday evenings in the summer of 1948. Elaine Noble recalls how her family and friends crammed themselves and a supply of "blankets, lawn chairs, brown paper bags of home-popped corn and a big thermos of strawberry Kool-Aid" into the 1940 Packard and arrived in time to claim viewing spots before the display window, which held four twelve-inch Crosley sets. "Promptly at 6:15," she says, "a gentleman wearing white trousers and a red blazer with a gold Gambles logo on the pocket stepped through the curtains at the side of the display." Ceremoniously turning the sets to face the viewers, the host tuned each to the test pattern and then "turned, bowed and retreated from view through the curtains to another round of applause." During KRNT's 7:00 cowboy movie no one in the crowd said a word, "not even during the commercials," and at 8:30, when the movie was finished, "the window went black," but the viewers kept their places for a while, taking their time to absorb the experience.[6]

Having enjoyed a freely roaming childhood in High Point, North Carolina, Harold Woodell says,

I first saw a television set in the window of a furniture store on the main street of my hometown in the early '50s. My brother and I rode our bikes in the evening to view this curiosity and found a crowd gathered watching the television through the window. There was no sound, of course, but it was fascinating nevertheless. At the time, I thought that television sets must be available only to the very wealthy, so imagine my surprise when I came home from school one afternoon a few weeks later to discover a set in the living room of our home. I was hooked and believe that I have now watched television almost every day for fifty years.

Bill Koon, who grew up in Columbia, South Carolina, finds interesting twists in his family's early viewing history:

We were a big radio family when I was a kid; my father was fascinated with any kind of technology, so we had plenty of radios. But especially we had a big one in the living room which we actually watched while we listened. So it was a bit ironic when we first saw TV, in that the first set I ever saw was in a furniture store window. And after my father would come home from work and we had dinner, he would load us up and we would go stand on the sidewalk and watch TV, which we could not hear. Maybe that says something about the tradeoffs in technological progress.

The second experience with the tube involved a daffy old uncle in Spartanburg who always had the first of everything. I remember driving two hours there to see (and hear) TV, but the signal was not clear that day, and we had to be content to be told that *Howdy Doody* was on, even if we could neither see nor hear it.

Anticipating an engineering career, Otha H. Vaughan, Jr., literally went to great heights to get an early glimpse of a television picture. When Atlanta's WSB was "testing their capabilities for transmitting TV in the spring or fall of 1948," Vaughan and his father lugged a small Admiral set with a round six-inch video screen to the top of a fire tower at South Union (near Seneca), South Carolina. "The reception was somewhat snowy, but you could still make out what was being transmitted [even though] we had a rather poor antenna. As a matter of fact I still have that TV in storage. My dad was a radio and TV experimenter who later opened his own repair business. He had built the first radio receiver in Seneca around 1923 and continued to experiment with radio and TV for the rest of his life."

By contrast, Jim Snyder first knew television in tentative and somewhat frustrating steps:

I grew up in northern Michigan, and it was years before TV reached our part of the world. I do remember that our church youth group was invited to the home of people who did have a TV with a very high tower which, they said, would bring in a Milwaukee station on the other side of Lake Michigan. The whole evening we sat enthralled looking at a screen full of snow (no sound). We did all think that for about five seconds we could make out the face of a clown. That was it.

My second experience relating to TV was in my freshman English class in college. The instructor assigned a paper on some specific aspect of TV—I've forgotten what. Well, I had never seen a TV program but did the best I could with the assigned subject. The paper came back with a very low grade and the comment that I should only write about things with which I was knowledgeable. I considered that comment very unfair, considering the assignment. Now that I think about it, I still do.

Although my fraternity house had a TV room, I don't think I ever watched more than a couple episodes of *Dragnet* in the whole four years. I don't remember that room ever being used very much. I guess there were just too many other things on campus, and TV programming certainly wasn't any great shakes.

Snyder, having formed an early preference for radio, did a stint as the editor of the North American Radio Archives publication *NARA News: A Journal of Vintage Radio.*

Jim Snyder's experiences parallel those of Jack French, another prominent figure in "golden age radio" interests today:

In the late '40s only a handful of professional people bought television sets in Rhinelander, Wisconsin, a little town not that far from the Michigan border and umpteen hundred miles from the nearest TV studio in Milwaukee. The set was basically just a piece of furniture, since there was virtually no reception. But if you were a doctor or a dentist, you were expected to have one in your living room. None of my circle of friends had a TV set then, so all I knew about this new entertainment medium was what I read in magazines.

My high school debate team made a trip to the University of St. Paul for an invitational tournament in 1951, and it was to be the first time I saw television. Shortly after our arrival on that Minnesota campus, one of our team came rushing back to us, eyes wide, and proclaimed, "There's a television set in the student lounge, and we can watch it!" The rest of us followed our intrepid co-member to a large, dark but friendly basement room, filled with at least 200 college students, augmented by the high school debaters who had drifted in.

At the very front of this enormous room was a television set, its tiny, round, bright screen filled with "snow" concealing all semblance of figures in an undecipherable program. Other than the snow flutter, nothing could be seen. (And believe me, coming from northern Wisconsin, I know "snow" when I see it.) The sound emanating from the TV set was too garbled and scratchy to provide a clue as to what type of program we were supposed to be watching. My debate team and I watched in the dark for about 15 minutes, saw nothing but snow, and could not figure out what was holding the rapt attention of all these students.

I finally announced, "This is stupid. We can't see anything. This will never replace radio." The team agreed, and we left in disgust, most of us under the assumption that we would never bother to watch TV again.

Reese Fant, on the other hand, saw promise even in his first tentative acquaintance: "I will never forget the first time I saw television. I was with my parents in the 'front room' of Harry Agnew's house, as we all crouched around his new television. It had a 7-inch screen and enough snow so that we were not only entertained, but we had contests as to what we were watching. We were, after all, trying to pick up a station in Charlotte, NC, with what is now called 'rabbit ears.'"[7]

Albert Holt, fond of quoting Keats, is sure that he "never got addicted to TV," but his first glimpses were intriguing: "I do remember that the early pictures on the small screen seemed like 'A new planet swims into his ken.' For awhile we looked at anything that materialized—from *Howdy Doody* to the Gillette boxing matches."

Brian Belanger's earliest TV-watching ambitions centered on early telecasts of the baseball World Series, and he went to some trouble to satisfy his desire to be an electronic spectator in Eveleth, a mining town in northeastern Minnesota:

I distinctly remember the first time I saw television. I was in grade school, and the year must have been 1950 or '51. In the early 1950s, the closest television stations were in Minneapolis and St. Paul, a couple of hundred miles south of us. No one in town had a television set, but we had all seen pictures of TV sets in national magazines, and we eagerly awaited the day when we too would have access to this new medium. The fall of that year the Elks Club in town decided to buy a TV set. They probably reasoned that the cost of the installation would more than be compensated for by increased bar sales, which proved to be true in short order.

The club was on the second floor of a two-story building. They erected a huge guyed antenna on the roof with a booster mounted on the tower. Since Eveleth was far out of the fringe reception area of the Minneapolis stations, the chances of having good reception were about nil. The decision to buy a set was probably based on the hope that the Minneapolis stations might be increasing power and that a new station would be coming on the air before long in Duluth, only about 60 miles away.

My father, a member of the Elks Club, told me that they planned to have the set installed and operating for the opening game of the World Series. As an avid baseball fan, I found this too much to resist. My elementary school was about six blocks from the Elks Club. We had an hour for lunch each day. There was no cafeteria, and so we kids walked home for lunch. My dad, along with every other Elks Club member who could get time off from work, intended to be there to see the series. Kids were not allowed in the bar room unless accompanied by an adult member. Dad agreed that I could join him.

The first day of the series, my mother packed me a peanut butter sandwich and apple lunch. I don't think I paid much attention to the lessons that morning. When the lunch bell rang, I flew out of my seat and ran as fast as I could to the Elks Club. When I reached the top of the stairs, the room was jammed—standing room only. The room was blue with cigarette and cigar smoke. The bar was doing a land-office business.

Much of the time the sound could be heard, but the picture quality was terrible. Often there was nothing but snow. Now and then a shaky picture would emerge out of the snow, and people would shout, "Look, there it is!" Even the people standing near the set could not see much on the tiny screen. For those at the back of the room, watching was a lost cause. Nevertheless, it was exciting. People were optimistic that, given more time, reception would improve.

I returned every day of the series, and every day I arrived back at school breathless from having run all the way to maximize the amount of time I could spend at the Elks.

After seeing television at the Elks Club, a handful of "early adopter" families in town bought TV sets. Once the Duluth station came on, others rushed to jump on the bandwagon. The fellow in Eveleth who sold television sets and installed antennas, Harry "TV" Keenan, had a backlog of orders during that era. You had to get your name on his waiting list to get your Trio Zig-Zag antenna and crank-up tower. The downside of being the first on your block to get a TV was that you would have a constant stream of neighbors dropping in for a visit to watch it—no privacy at all.

My family did not get a television set until 1955. We were one of the last families on our block to do so, much to my consternation. Of course we had to buy a set of TV trays so we could eat our evening meal in front of the set lest we miss anything. Eventually we became more critical of program content and more selective in our viewing, but for the first few weeks the novelty was so great that we would watch whatever happened to be on.

Today I find relatively little on television that I feel compelled to watch. I find it rather quaint that I could have been that excited about it 50 years ago.

In the early months of 1953, "the Singing Rage, Miss Patti Page" sang Bob Merrill's hit parade song which asked in lilting anxiety, "How Much Is That Doggie in the Window?" For most citizens, however, the more burning questions of the day seemed to be "How much is that TV in the window?" and "How soon can I take it home?" and "Do you have a time payment plan?"

Notes

1. Quoted in Ray Barfield, *Listening to Radio, 1920–1950* (Westport, CT and London: Praeger, 1996), 31.

2. Interview segment, "Television: Window to the World," *Modern Marvels*, The History Channel, first broadcast 1999; repeated September 3, 2003.

3. Jeff Miller, "A U. S. Television Chronology, 1875–1970," http://members.aol.com/jeff560/chronotv.html, accessed August 12, 2003.

4. Jay S. Harris, with *TV Guide* editors, *TV Guide: The First 25 Years* (New York: Simon and Schuster, 1978), 36.

5. "Kukla, Fran & Ollie," *Museum of Broadcast Communications Encyclopedia of Television*, ed. Horace Newcomb, vol. 2 (Chicago and London: Fitzroy Dearborn, 1997), 910–911.

6. Elaine Noble, "When 'Window Movies' Came to Des Moines," *Reminisce* (July-August 2003): 12–13.

7. Reese Fant, "If You've Got Any Radio Stories, There's Someone Who Wants Them," *Greenville News* (September 19, 1993): sec. D.

$$\text{---}\ 2\ \text{---}$$

Test Pattern Days: "We've Got a TV!"

Mary Steed remembers the time about fifty years ago when her family purchased the first television set in her Brooklyn neighborhood. The delivery van pulled into the driveway, the installers ceremoniously carried the large cabinet into the house, and they spent a long time adjusting knobs to the midafternoon test pattern. When word spread in the neighborhood and at school that her house harbored a TV set, she soon noticed a change in her social life: "My dating activity suddenly rose to a zenith. When other homes got their TVs, my dating scaled back to normal." Shakespeare scholar Albert Holt discovered a different benefit of TV ownership: "TV early became the most used baby-sitter and made hiring baby-sitters much easier." Linda Law remembers when her family bought a TV receiver in 1951: it was "Pandora's (cathode ray tube) box for the 20th century! Black and white, of course. So many images come back to me now: Howdy Doody, Ed Sullivan, Milton Berle, *I Love Lucy*, Burns and Allen, Sgt. Bilko, and late at night, pro wrestling (which I wasn't allowed to watch, but my younger cousin, Pam, was!)" She recalls too the repeated parental admonition "Don't sit so close to the screen," the suspicion being that the new medium might irradiate or blind its viewers even while it informed and entertained them.

Sitting close to the screen was a natural temptation when picture tubes measured under ten inches. To make the picture seeable, the owner of a very early set might purchase a large magnifying lens that sat on a stand between the receiver and the viewer. According to Owens Pomeroy, "In those early days, before the big screen TVs came into view, some manufacturers offered an enlargement of your 7-, 10-, or 12-inch TV. It was a glass globe, flat on one side and filled with water or mineral oil, that you placed against the screen. The only drawback was that you had to sit almost directly in front of the screen or else the picture would be distorted. This was short-lived when DuMont came out with the first 16-inch picture." As Mary Lynn Moon recalls, "We had the first TV in our neighborhood. I don't know how we afforded it, but my father was always intrigued by electronic products. It was a 7-inch, but it had a blue 'bubble' attached to the front to enlarge the picture. We lived in a very small bungalow, so when the neighbor men came to watch a program (my memory is of boxing), my father would turn the TV to the window, turn the sound up, and they would sit on the porch to watch."

The Samuel Green family of Worcester, Massachusetts, also found that guests came with the TV set, as daughter Myrna McKee recounts:

> It was late in 1945, just after World War II, when Daddy brought home a new appliance. He told us it was called a "television." This new toy came in a big dark brown wooden box with a little 8-inch screen in its center.
>
> Daddy placed it carefully in the corner of the living room and plugged it into the wall socket. He turned the "on" button with a click, and slowly the screen came to life. Large black and white lines made a design with the numeral 4 in the center. It was our first test pattern.
>
> I had heard all about the new medium of TV, but to have one sitting in our living room was thrilling. Later that evening several close friends of my parents came to our home to view this new addition. This started a tradition. Every evening two couples would arrive in time for the evening programs to come on. The living room began to resemble a theater. Chairs were set in rows and Mama always made popcorn. . . .
>
> Time passed and my parents' friends bought their own sets, so the evening theater stopped at our house. Mama was glad to get her living room back. The test pattern was shown less and less as more programs filled the airways. *I Love Lucy* came along, but still Milton Berle owned Tuesday night. He was first in the ratings and now was called "Mister Television." Daddy bought a rotary antenna and it stopped snowing in the house.[1]

Neighborhoods and communities made similar adjustments to the arrival of television sets. Historian Doris Kearns Goodwin told interviewer Leah K. Glasheen that in the 1950s Long Island community of her childhood, "The addition of a TV set in the neighborhood became a public event, . . . adding another stop in the parade from one house to another to watch particular shows. (*Howdy Doody* was the Kearns family specialty.")[2]

In the post–World War II economy many returning service men and women were concerned with completing their formal education or trades training on the GI Bill, starting or expanding businesses, and buying houses for their young families. Dan Galvin says that his friend Gene Zaner's brother "returned from the war in 1946 with some accumulated back pay (traditional for servicemen on active duty for several years) totaling the then-lavish sum of $400. In the Zaner family's socioeconomic milieu of Bayhill Road (mostly Italians and Jews) this was a princely sum, just enough to buy a 10-inch screen Philco of 18 inches high, 24 inches wide, and 18 inches deep. Since no one else in the Brooklyn neighborhood had a television set, the 'block' presented itself each Friday evening to be entertained by Buffalo Bob, Howdy Doody, and the Great Foudini (a pseudomagician embroiled in constant pratfalls)."

Jerry van de Erve figured his family budget closely in those days. Promoted to the position of Eastern Area Engineer for his government agency, he moved his family from Seattle to Washington, DC, and prepared to travel the entire East Coast, as far west as Chicago. He says,

The children came down with the measles just as I had started my first field trip. To help my wife occupy the children in a new place without my help, we bought our first TV. The key men at my headquarters all told me the set I should buy was a 12-1/2-inch black and white Philco table model TV for $250. That was no small amount. I had a good job at $2,600 per year. We paid $10,000 for a three-bedroom, 1-1/2 bath brick house with slate shingle roof in Bethesda, an excellent suburb. Next to an undeveloped state park we had 21 dogwood trees in our large front yard.

In the fall of 1953 we were transferred to Sacramento, California. It is now about the twentieth U.S. city in size, but in 1953 there were no TV stations in Sacramento. We paid $19,000 for a three-bedroom, two-bath house in the best part of Sacramento. An added cost to the house was ... $100 or so for the 40 foot high TV antenna on the roof, tall enough so we could tune in two or three TV stations in San Francisco, about 100 miles away.

Another viewer reports, "There was always a television in my life, even though there wasn't always a television signal." Here too, a family move made a difference, as he explains:

The war was over, and my young parents bought their first television set in 1949. Their first-born was only four months old, and the new "stay-at-home" lifestyle justified the purchase. It was a General Electric model. And it wasn't the standard 8-inch model, or the 10-inch, but the super, "jumbo-wide" 12-inch screen.

The number of television stations in their Michigan viewing area was three. But when they moved the family of four to Summerville, South Carolina, there were none, and the unit spent the next two years in the corner. I was two years old when the Charleston television station started broadcasting. I remember at that early age how life was in a small Southern town in 1953. Most of the roads were dirt—some horses and wagons were still being used in town—and air conditioning didn't come to our household for a few more years.

In 1953, Charleston's WCSC came on the air, but regular programming did not begin until about noon. When programming was not on the air, a test pattern was present. The test pattern I remember was of a Native American Indian with full-feathered headdress in the screen center and radiating patterns outward. The new technology was not surprising to a two-year-old, but for a teenager living in a sleepy Southern village, it must have been miraculous. My very earliest memories of this "first TV in town" were of my teenage babysitter. While my parents were at church on Sunday mornings, our babysitter would spend most of the hour and a half sitting in front of the set ... staring at the test pattern. As far as I know, she never saw any actual programming since her sitting job was completed by noon.

Just as a firstborn newly brought home from the hospital becomes the center of family attention and causes rearrangements of furniture and schedules, the TV set made its demands, and "care and feeding" issues arose very quickly. Advertisements for eye drops cautioned against the dread condition called "TV eye"—easily incurred by watching the entire four- or five-hour broadcast day of the pioneering TV station. Word spread that TV should be watched in subdued light, as Richard Leiby was warned when he had his "first TV experience" of watching "a 9-inch round black and white set in a totally dark room" soon after his college graduation in 1949. Owens Pomeroy points out that the room lighting misconception arose when "TV was only on the air from about 7 P.M. to midnight, and we thought you had to view it (for a clear picture) in the dark, like home movies." When stations began to expand their telecast days into the morning and early afternoon hours, he notes, "They even had commercials saying TV is just as good and clear in the daytime as it is at night, and you do not need to view in the dark."

Frank Clipp sold TVs in his New Albany, Indiana appliance store in the late 1940s. "I knew nothing about 'em, but I sold 'em," he says. "The first time we demonstrated them to an audience, we got about 50 or 60 chairs from the funeral director and turned out all the lights because we thought it was like a motion picture, you know, and watched the high school football game from Louisville, Kentucky. It was terrible." He had little trouble with the sets he sold—luckily so, for he had no TV service man—but in New Jersey he had earlier experienced a problem with the first set he bought for his own family: "A tube went out the first night I had it, and my kids raised Cain."

Furniture placement figures into Elizabeth M. Blakely's account of her family's early viewing history:

TV in the '60s for me meant a B&W TV with rabbit ears that had to constantly be adjusted. The TV was placed on a window seat at one end of our living room that also served as my dad's study. There was only one chair at that end of the room (the term "couch potato" had no meaning then) so my older siblings and I always tried to get that seat! Otherwise we'd sit on the floor and crane our necks upward to watch TV.

I remember that in the late '60s we got a small 12 or 13-inch color TV (GE-made, which is where my dad worked). I'll never forget the time my uncle visited and he and Dad put the two TVs next to each other so that they could watch two college football games (probably bowl games) at the same time. (The color TV deteriorated over the years to provide predominantly orange and green colors. It was a challenge to see the ball for any sporting events!)

We moved in the early '70s, and our new home included a den. So we put the TV in the den, and a couch was purchased so that more of us could watch TV in comfort (though there was still no remote). My father had his desk and chair in the den as well and would often swivel his chair around to watch TV at night with us kids (*Quincy, Adam-12, Baa Baa Blacksheep, M*A*S*H, Charlie's Angels, The Partridge Family*), and answer homework questions during commercials. My father would think about my math problems while staring off into space. I would await his "aha" while watching the evening programming (which was such a luxury to watch with all the lab reports and homework we had).

The dinner table arrangement in Mary Lynn Moon's childhood home put her father's back toward the television set. His solution, she says, was to watch at mealtimes "with a mirror stuck in a milk bottle" strategically placed on the table. By contrast, sociologist Cindy Roper observes, "When my brothers and I were young, we had one TV for the whole family, and we would never have considered watching it while we ate dinner. Now we have two TVs in a two-person household. My brother, who has two daughters, has four TVs in one household, and these have sophisticated satellite and recording capabilities. So I think one of the most striking differences between now and then is not the language and content, which have certainly changed, often for the worse, but how we view TV."

First-TV-generation questions of purchase price, adequate reception, and program quality caused many families to wait for months or even years before purchasing a set, but Caroline Whitmire Todd's family had the interesting dilemma of owning a receiver for some time before programming was available: "When my Uncle Jack [Brennen] died in Atlanta, we inherited his DuMont console television. Because we got it before television had come to our home in Columbia, South Carolina, we could only get the test pattern that came on every hour on the hour, and maybe on the half hour. My brothers John and Bohler and I would run into the living room, turn on the TV set, and sit in front of it, watching nothing but the Indian on the test pattern. Sometimes we invited our friends over to watch it, too. We all loved it so much that I now remember the test pattern much better than I remember any of the programs we watched when TV came to Columbia."

In Iowa, Marilyn Chadwick tells the story of her introduction to television with typical enthusiasm:

I was approximately nine years old when television entered upon the scene. Family evening activities had thus far centered around the radio, where we gathered for such shows as *Corliss Archer, Judy Canova,* and *The Shadow.*

My imagination could scarcely comprehend the new invention that my parents described to me . . . an actual picture screen that could be purchased and viewed in the home of anyone who was able to afford such a device!

All those early televisions were either consoles or what was called the table model, which was far from being portable. Those picture tubes were incredibly large and very heavy. When the picture tube bit the dust and was beyond repair, its owners were reluctant to part with what was termed a "nice piece of furniture." Therefore, many a console, after undergoing a "tube-ectomy," was recycled into a new function as perhaps a liquor cabinet and relocated to a knotty pine basement rec room.

Our family of four (plus dog) was most fortunate in being the first in our neighborhood to have our very own TV. We were advised that this purchase was the cream of the crop and top of the line. It was a Sylvania with a "halo light." This light framed the picture on all sides and was pitched to the consumer as being easier on the eyes than a model not so endowed. We had to forego the rooftop antenna, probably because of the extra cost for the halo! We settled for second best in the reception department, which was a set of rabbit ears, a.k.a. bunny ears. These worked quite nicely, and since we only had one channel, it was seldom necessary to change the position of the ears, although doing so gave one the feeling of immeasurable power. If there was some strange interference to reception, Mom would take emergency measures by wrapping aluminum foil around the tips of the ears. I was never interested in an explanation as to why this worked!

In Brenda Seabrooke's family, the passion to have a television set struck the father, not the child. On a driving vacation trip from Georgia to Washington, DC, ten-year-old Brenda and her parents saw their first television program at a motel in Springs Lake, Virginia, according to her diary entry for July 10, 1951, a day when they also visited Martha Washington's family home. She explains,

We had only heard of television in my hometown of Fitzgerald, Georgia; nobody had actually seen it. After this first taste, my father developed a great desire for a television. One night after supper he, my mother, and I drove all over town looking for a television he'd heard about on the north side. In those days you couldn't conceal having a television set because to receive programming, you had to have an aerial on the roof that reached halfway to Heaven. We never found the rumored set, but a few days later my parents went to Atlanta, a distance of 180 miles. I think my father must have had to order a television because he returned to Atlanta a week later and came home with a Philco cabinet model in the backseat of his car. (It wouldn't fit in the trunk with his suitcase.) As far as I know, it was the first TV set in town.

While we watched TV a lot, it didn't seem to take over our lives. This may be due to the lack of channels—there was only one—and to the awfulness of the reception. The snow on the black and white TV was the first snow of any kind I had ever seen, and it was hard on the eyes. Sometimes we couldn't even tell what was happening on the screen. There was no remedy for the rolling image which persisted into the '60s, so that my son learned as a toddler to whack the TV on the right side the way his grandmother did to try to stop the picture's rolling and waviness!

In his childhood Phillip White had to live with "a very dull TV," a circa-1958 model that was bland in appearance and uneager to produce a watchable picture:

The story that my grandfather loved to tell until the day he died was about our television reception problems. I was only two or three years old, and my grandparents would come to visit. My grandfather would inquire about what was on TV because he knew that as soon as the set warmed up, the screen would develop static, and the picture would become distorted, and I would laugh at the people walking crooked. Then my grandfather would instruct me to go "fix" the television. I would get up and go over and slap the right side of the cabinet as hard as I could, and the picture would straighten up for a while. This was the way that my parents had to watch television for quite some time.

Although my grandfather got a kick out of watching me "fix" the TV, I often wonder how our spoiled remote cable cruisers (especially myself) would be able to tolerate such a disturbance today.

In a South Carolina college town halfway between Atlanta and Charlotte, Joe Bailey and his family had sampled television in the early 1950s at the home of a neighbor, a retired chemistry professor, but the picture was never very good. In late 1954, however, the Baileys heard the encouraging news that WFBC-TV in Greenville, thirty miles away, was due to begin broadcasting "around the end of the year" and planned accordingly:

> Our family all chipped in together and bought a TV for Christmas, a 21-inch Admiral from Horace Crenshaw's store in Pendleton. They put a big pole and antenna on our roof, and my brother and I went up on the roof with a compass and aviation map and very carefully lined the antenna up on Greenville. Of course there was nothing [yet] to watch but very snowy images from Atlanta and Charlotte. About that time the Greenville station started broadcasting a test pattern, so we would watch that test pattern and marvel at how clear the picture was and would adjust every possible control on the TV to make it as perfect as possible.
>
> Except for Dr. Pollard, we had about the first TV in the neighborhood, and I do remember that during the bowl games we had a house full of neighbors watching with us.

When the Bailey family and others viewed the inaugural telecast of Greenville's Channel 4, they were startled to see that, after introductory remarks from civic and station officials, the first movie to be shown on the station appeared upside down. "That was nothing," the station's chief engineer says; "spilled reels of film, projector failures, falling mikes—it happened all the time."

In Encinitas, California, Jim Bivens built his own first TV set from an RCA kit. He disapproved of the fact that the condensers were made of cardboard, so he prevailed upon RCA to sell him ceramic ones. His patient wife Hazel "allowed him to fool around with it, but lived in terror that he would set fire to the house with 'all the loose electricity,' for she was under the impression it took high voltage to run the TV, and it frightened her to death." By the time the set was completed and ready to be tested, according to the recounting of a friend of theirs, both husband and wife had become wary because of "all these scare stories about high voltage coming out of the back of the tube," so they asked an electrical engineer friend "to come over and stand by. He did . . . the set was turned on . . . it worked perfectly— probably, Jim says, the only time that it worked perfectly through the remaining time they had it!" Subsequently the Bivenses seemed to be victimized by "a spotty signal that resulted in poor reception—maybe it was the set—who could tell? Poor picture or not, they still became a neighborhood center for watching the most popular regular program, the wrestling matches."[3]

Clark Wickliffe recalls how his father, "a tinkerer," ordered and assembled a Heathkit TV and built a cabinet for it, and he "got behind it every morning and fine-tuned the WFBC-TV Indian head test pattern reflected in a mirror placed against a chair." When the elder Mr. Wickliffe, a high school principal who could silence a rowdy basketball crowd by simply standing and facing the bleachers, was invited to be interviewed on the station's local morning show, he arrived while the test pattern was still being shown and was given a tour of the studios before the day's programming began. When he met the engineers, he pointed out that the test pattern being transmitted was not properly aligned, and, his son remembers, he "chewed out the engineers for their sloppy presentation of the test pattern that made his tuning job harder."

A family crisis brought TV into John Dodge's life in 1949, when he was six years old. His mother had been hospitalized with a severe mental illness, and his father was time-strapped by his film industry public relations duties. John's care fell to his seventy-two-year-old grandparents and to his aunt, and during the week he attended "an exclusive private school where I boarded. I came home weekends, holidays, and summertime." Radio was the main entertainment at school, although the students were allowed one hour of TV each evening, enough time to catch *The Lone Ranger* on Thursdays on the Los Angeles ABC channel and a musical program hosted by the exotically turbaned Korla Pandit on KTLA, "historically the first TV station West of the Mississippi," according to his vivid memory for such details.

John Dodge's young appetite for television grew when he came home one weekend and found that

> my grandfather had just purchased a 16-inch Motorola which was brown and had rotary knobs. On-off and volume [controls] were on the left and the station tuner on the right.
>
> I dominated that TV! Friday nights I always watched *Western Varieties* with Doye O'Dell, and Saturdays between noon and 3 P.M. I watched the great B Westerns, which later in life I devoted time to collecting on videos. On Sundays between 3 and 6 P.M. was the greatest lineup a little boy ever saw! From 3 to 4:30 was Hopalong Cassidy. From 4:30 to 5:30 was the East Side Kids, and from 6 to 6:30 was a chapter of *Flash Gordon* with Buster Crabbe. All those weekend shows were on one station, KTLA, Channel 5.
>
> Were there any technical problems? We had an outside antenna on the roof, and I remember the brown flat rubber wire which went from the back of the TV along the floor, out a screened window, and up to the roof. The only problem that developed was that I would turn the station tuner knob too fast and periodically we could not get some stations clearly and over came the TV man, who told my grandfather that little John was being too rough on the tuner knob. The tuner would need cleaning and adjusting.

Just as a farmer's mind must always be monitoring the weather, the owner of the pretransistor television set had to worry about tubes. Each turn of the "on" switch was an opportunity for a crucial tube to die in a sudden burst of blue-white light or to fade from its usual bright orange glow to a silver and gray-brown demise. While some owners obeyed manufacturers' injunctions against venturing beyond the perforated back panel, others regularly lifted out suspiciously dim tubes and rushed them to the testing machine at the 7-11 convenience store or the pharmacy. Thus a 1954 Perry Barlow cartoon for *Collier's* shows the TV service-call repairman reaching solemnly into the back of the set while the owner wears a "this is going to cost me plenty" look behind the pipe clamped in his teeth. His daughter innocently suggests in the cutline, "Tell him about the pretty blue flash when you fixed it."[4]

Owens Pomeroy, first in his block to have a set, looked into the back of the twelve-inch-screen Crosley floor model and marveled: "That little son of a gun had no less than 50 tubes and felt like it weighed a ton!" For Walter P. Horlyck in 1950s San Francisco, the problem came, more than once, with the biggest and most expensive tube—the picture tube:

One weekend the San Francisco Seals were playing Los Angeles in a BIG series starting Friday night, and since it was in LA, it was to be televised. Friday night, in great anticipation we finished dinner. I opened a beer and went in to watch the game. Turned on the TV. NO PICTURE!!! No sound, no nothing. Same all weekend, so being the astute repairman, I diagnosed the problem as a tuner problem. Monday morning I removed the chassis from the set. In those days everything came out as a unit, picture tube and all. As I'm going out the door, the dog and I arrive at the very same instant, and I proceed to trip over the dog. In order to keep from dropping the chassis, I swung it around and hit the neck of the picture tube, and the implosion made a big sucking sound as the neck broke off.

With a few choice words, I put it in the back of the car and hauled it in to the Sears service center, where I worked in appliance repair. I put it on my buddy's bench and waited for him to come in. His first words were "What the hell happened?" I explained the obvious, and he said, "What's the problem, other than the picture tube?" I explained that I thought I had a tuner problem, at which time he started to laugh. With a few choice words I said, "What's so damn funny?" He said, "That station went off the air last week!" About that time the shop supervisor came in and said, "Walt, you're in luck. We have a picture tube with a small bubble in it, and we can't use it for a customer, and you can have it for five bucks." So they put it on, and I took the chassis home. Since my wife wanted to go shopping, I just put the chassis in and didn't secure it or put the back on the set. The TV sat in front of some floor-length windows with [partitions] running from the floor to the ceiling. As I'm getting dressed, I hear this sucking noise, just like this morning. My wife comes in and says, "Didn't you put the back on the TV?" My heart sank. She said, "Well, I thought it was done, so I pushed it back to the windows." Sick at heart, I went out, and sure enough the neck to the CRT [cathode ray tube] was just hanging there AGAIN! "Well," I said, "that does it. We just won't watch TV again, ever." After a day of I mean DEAD silence, I agreed to take it back to the shop. I would've rather eaten a big green worm than to show up again. I left home very early to take it in and put it on the bench. My buddy came in and said, "Walt, what the hell did you do now?" The shop supervisor came in and said, "If I knew you were going to do this, I'd have kept the CRT for myself." I told him, "If I knew you were going to do that, I wouldn't have done it." I then said, "Go ahead and put another CRT in it and have a TV tech take it back." He said, "I'll have to charge full price for the tube." An expensive repair. From then on the big joke in the shop was if you have a TV tube that has nothing wrong, let Walt work on it, and there will be plenty wrong when he gets done.

Since that day I won't go near a TV that doesn't have a back on it. The moral to the story is "Leave well enough alone."

Robert Cox, who "thought it was a big deal to sit up 'til midnight and listen to the national anthem and watch the station sign off the air," recalls that his father "was never quite satisfied with the quality of the picture on our Silvertone TV, always jumping up from his easy chair to make adjustments with the knobs, making the picture lighter or darker and fooling with the vertical and horizontal while my mom complained, 'Leave it alone. It was all right before you started messing with it'—a scene that was probably acted out in many living rooms all across America."

In the first television set that Cindy Roper knew ("a huge black and white job that sat on a TV stand, with two knobs, one on each side, for volume and channels"), a small door in the center below the picture tube "flipped down for adjustments (which, as kids, we weren't supposed to do)." Other sets had horizontal, vertical, contrast, and brightness knobs temptingly arrayed in the back of the cabinet. If the program choices were uninteresting or the reception hopeless, children sometimes amused themselves with unauthorized "scramble the picture" games.

Other children simply longed for a set to watch. "My father would occasionally bring a TV home just on a trial basis, so we didn't see it often at our house," says Ohioan Mike Newton. "There was a neighbor child who had the only television on the block, so naturally we treated her REALLY NICE. Anything she wanted to play, we played. That's so we could go over to her house and watch *Howdy Doody!* Although there was not always a TV set at home, Mike Newton and his sister had their own television seats: "My mother painted two small chairs for my sister and me that were really too little for us, but she put our favorite shows on them in the shape of hearts. Unfortunately she later painted over them. I have since seen TV chairs from that period with names like Hopalong Cassidy or Howdy Doody."

Many a TV-denied child of Priscilla Kanet's generation would empathize with her sigh as she conjures images of her young days in rural Maryland: "I think my family was the last one on earth to get a TV. My dad finally broke down and bought one of those 'frivolous things' because we wouldn't stay out of the neighbor's house. TV became part of our play—lots of fun cowboy stuff." Television was not a uniformly welcome addition to Joan W. Henry's childhood home either: "My father thought television was a waste of time. He believed in reading and writing and looking up words in dictionaries. So, of course, I used to sneak down to a neighbor's to watch *The Mickey Mouse Club.* Years later my father relented and bought a small black and white. My poor deprived brother (fourth grade), who was a periodic sleepwalker, would come out into the living room in the evenings, sound asleep, and pat the new TV on its head—just to make sure it was still there, I suppose. My father ended up getting much pleasure from television during his later years. I don't like it much and prefer reading and writing and looking up words in dictionaries."

"I must tell you the story of how our family got our first TV," Sara Robertson says:

It was about 1957, I think. We were rural South Carolinians, poor as the proverbial church mice. My daddy worked in the cotton mill and farmed. Our mother had died a couple of years before, leaving Daddy with five children at home and twins at Thornwell Home for Children. All of our friends and cousins had the wonderful new TV, and of course we had seen theirs and yearned for one of our own. All of a sudden the AM radio was just old hat. Anyway, we kept pestering Daddy for a TV, and finally, to hush our whining, he told us, "When Aunt Kate gets a TV, I'll get you one." Well, Aunt Kate was his elderly aunt, who was notoriously tight with a penny. She was forever talking about how "high" everything was and what a shame to waste money on this or that. That seemed to be the end of our pleas for a TV. We occasionally visited Aunt Kate and Uncle Ed, who lived over in Anderson County. Daddy would load up all the kids in the car on a Sunday afternoon and we would go to see one relative or another. Not too long after this pronouncement, we went to visit Aunt Kate, and, lo and behold! There sat Aunt Kate in front of a shiny new television set! It wasn't long until we had one of our own.

For Jane Walker's family in Chattanooga in 1957, the TV purchase issue was a multifaceted dilemma:

We put off getting TV when most of our neighbors had it because we felt it was better that our sons do other activities than just being TV watchers. We were disgusted that the boys' Cub Scout Den Mother would have the boys come to her house and would just have them watch TV. I finally volunteered to be the Den Mother and involve the boys in real scouting, and I think she was happy.

Also at that time, one had to have an aerial over the house, and there had been robberies at homes that had TV aerials—another reason we delayed.

Finally, though, we consented to our sons' frequent requests to get a set, and we will never forget the date of that very first program they saw. All of us remember. Our beloved Irish setter (Bonnie Sandra of Bay Ridge) was 14 years old and very ill, having to be fed intravenously. The boys were watching *Science Fiction Theatre* that day when I went out to check on Sandy and found that she had gone to "dog heaven." Of course I had to come in and tell the boys, and they never did get to see the end of their very first TV program.

Some families came to TV set ownership with all slow deliberateness, while for others the TV purchase created a Zone of No Resistance. "The picture was not clear and the sound was fuzzy," says Jean Jackson of her 1957 black and white TV initiation, "but we watched it anyway." Nancy Carls' family preferred the cautious approach:

> I think it's important to note that when I was young, we had no TV and really did NOT miss having a TV—except maybe when *Bandstand* came on . . . ha. I always went to my friend's house to watch *Bandstand* where even her mother would sit down and watch.
>
> When I was a kid, we played outdoors a lot, whereas kids these days sit in front of the "boob tube" and don't get that enjoyment of playing outdoor games like "Red Light/Green Light," hopscotch, jacks, and all those other games you don't hear about these days!
>
> It seems like I was around 11 or 12 when we finally got TV. I must admit it was exciting. I recall making it a tradition on Saturday nights to sit down in the living room with my folks, sipping cocoa, eating Archway cookies, and watching *Gunsmoke*. My folks always watched Ed Sullivan's show and of course Lawrence Welk (their favorite). I personally liked the Ozzie Nelson show because I had a major crush on Ricky Nelson.

On the other hand, Allan King's family in New York City simply took the plunge. Already the city had an abundance of program and channel choices:

> We got our first television in 1955, and I don't remember its ever being off except for those times when the stations were off the air and nothing was being televised except static or a silent test pattern.
>
> The adults favored the early comedies—*My Little Margie, December Bride, Pete and Gladys, Oh! Susanna, The Ann Sothern Show, Topper, Amos 'n' Andy, The Jack Benny Show, I Love Lucy, I Married Joan, Ozzie and Harriet, The Donna Reed Show*, and so on—while we kids watched for *Howdy Doody, Captain Kangaroo, Kukla, Fran, and Ollie, Ruff 'n' Ready, Shari Lewis, Beanie and Cecil, Merrie Melodies Cartoons*, and so on. We all seemed to enjoy the adventure shows, of which there was an abundance, especially the Westerns. Then came the crime thrillers and detective shows. *M Squad, Racket Squad, Undercover Policewoman, Dragnet, Peter Gunn, Mr. Lucky, The Naked City, Boston Blackie*, and *Mike Hammer* come readily to memory. Add a few live variety shows, and early television was a tough act to beat.

Railroad man Royce Augustine Hoyle was so pleased to have a television set in his Savannah, Georgia, home that he took hundreds of still photographs of it in operation. According to his grandson, Walter Hill, between 1957 and about 1962 he kept his camera "always at arm's length" near his viewing chair and shot almost 700 pictures of the screen, always centered in the frame and surrounded by blackness. The grandfather demanded silence when the program was on, and "the dial would eventually stop on whatever station carried Lawrence Welk," producing "a collective, ever-so-quiet groan" from the grandchildren.[5] Mr. Hoyle died in 1969, and his wife's death in 1983 prompted the Hill brothers to look again at the photo albums on the shelf. Beyond the family and flower pictures they found a private record of programs seen: snapshots of quiz show contestants,

President Kennedy, Mitch Miller leading a singalong, Ed Sullivan and his guests, and so on. In 1988 the Hill grandsons decided that the pictures, beyond being a part of the family chronicle, had a place in media history, and they issued a booklet reproducing eighty-eight of the television images.

"We were TV snobs," Roger Rollin says of his young married years, and his family was finally persuaded to invest in a TV set because of a desire to follow the 1960 political conventions. Much earlier than that, Roger Rollin's educational pursuits, military service, and teaching responsibilities had made him a distant observer of the medium:

The first "real" television set I ever saw was in a Washington, Pennsylvania, hardware store window in September 1949. I don't remember what was showing on the tiny screen, but I do remember thinking, "Why, that's television." I wasn't impressed. I didn't stay to watch. I didn't ache to own a set of my own. Being in college at the time must have had something to do with it—there was just too much else exciting going on in my life for me to give much thought to television. I did not experience the excitement when the first TV set appeared on the block. I didn't try to wangle an invitation to watch a neighbor's set. And besides, even the little college town I lived in had four movie theatres, so who needed television?

Then, about a year later, my parents got a set. Since I often came home to Pittsburgh on weekends to leave off my laundry, borrow Dad's car, and go out on dates, I had a chance to view television on a more regular basis. I still didn't find it addictive. Even when, in 1951, my college fraternity purchased what was then a huge set (probably about 19 inches), watching TV was mainly a Sunday evening activity, a chance for all the guys to sit around and watch the Dave Garroway or Perry Como shows and make lewd and lascivious comments about the female singers thereon. I can't recall any of my fraternity brothers, even the most notorious layabouts, becoming what would come to be called "couch potatoes."

The Korean War began when I was in college, but it was not, like Viet Nam, a "television war." All I can recall seeing was the occasional combat newsreel. To learn about this "police action," we read the papers. I went into the Army after graduation, in 1952, and served for three years. I can't recall seeing a television set in a barracks, a PX, or an EM club between 1952 and 1955. The one exception: in 1953 a new comedy team called Martin and Lewis were beginning to receive a lot of attention. My wife, Marian, and I heard that they were going to appear on television on a certain evening, and we also heard that our Monterey, California, USO had a TV. We joined about a hundred GIs roaring at what then seemed to be inspired zaniness. But in 1954 and 1955 there was no television equivalent of the U.S. Armed Forces Radio Network in Austria and Germany, where I was stationed. I'm not even sure that German and Austrian national television even existed then, and no Germans I knew had a TV set.

Earlier television sets were expensive, and certainly more than I could afford as a graduate student. Even when I secured my first teaching position, at Franklin and Marshall College in Lancaster, Pennsylvania, and we at last had a home of our own (rented), money was still tight, and we felt no compelling need to own a TV set.

Politics led to our acquiring our first television set. I became intensely interested in the 1960 presidential race and mentioned to my wife that I'd love to be able to see the Democratic National Convention, adding that it was too bad we couldn't afford the $100-plus the cheapest TV set cost. (My annual salary as a college instructor was at the time $5000, and I was very glad to get it.) Marian replied that we could afford it and to my astonishment revealed that she had been saving small amounts of money for just such "emergencies" over the last several years. A friend told us that he knew where we could get a Zenith demo for $100, and we were in business.

As viewers spent more and more hours facing the tube, the remote control device became a welcome accessory, although the early ones were not wireless. Rather, the remote control cord, sometimes acting as a treacherous jump rope to unwary feet, was strung from the TV set, past the coffee table, to the arm of the La-Z- Boy lounger. Even with its tripping danger, the early "remote" offered the first primitive exercises in rapid channel selection and saved the labor of crossing the room and manually pushing or pulling the small on/off knob or turning the volume control on the front of the set.

Mitch Fields of Royal Oak, Michigan, well remembers the wireless Space Command remote that controlled the Zenith receiver at his grandparents' home in the 1960s: "We didn't use it for surfing—after all, there wasn't much to surf: there were only three broadcast channels, or four if you counted the Canadian station we got out of Windsor, Ontario. Mostly we used it for muting commercials. The aluminum rod technology did have its drawbacks: setting the table with silverware or jiggling a set of keys would generate sounds that would result in unpredictable channel changing, muting and powering off and on that we kids thought was very cool."[6]

Although print ads for early TV remotes associated them in one way or another with space-age wonder—a device that Captain Video might use for his own leisure watching back at the Space Barracks—a clever workman could fashion his own. Beth Jarrard speaks of her Uncle Russell, who "designed the first remote control I ever saw. He had little formal education, but he worked at a textile mill repairing machinery, and he had a gifted mind for all things mechanical. He rigged up a gizmo with an electrical cable from the TV with a little push button that advanced the channel selection. Of course, we had only three [station] choices at that time, but it was the advent of channel surfing in Marietta, South Carolina."

For many children of the 1950s, there was another "essential" TV accessory. Brenda Seabrooke understood that the transparent color strips that her father applied to the screen were primarily meant to reduce "the snow glare." Other young viewers, those living in closer proximity to station transmitters, knew a different purpose for the green, blue, and pink see-through plastic, although their recollections of the tints are somewhat varied. Dawn Galvin remembers watching "the store-bought green and red transparent overlay sheets fitted to the shape of the black and white screen" in her home "in the immigrant mix of the South Bronx (i.e., poor)." Owens Pomeroy wonders " . . . how many remember the 'color sheets' for black and white television that you attached to the front of your 7-, 10-, or 12-inch screen? It consisted of nothing more than a tinted sheet of plastic in three sizes, green at the bottom, flesh-tone in the middle, and light blue at the top. (We had one.) This was OK if you were watching a show that had a long shot of a city street or Western prairie, but if there was a closeup of a person, they had green clothes and blue hair! I think this idea lasted about one TV season (32 weeks)." R. Bernard Chapman, Jr., understood that the "see-through plastic sheet" was "supposed to give the impression that the black and white TV was a 'color' TV. What can I say? It worked for us." Harold Woodell was not so well pleased with the mail-order purchase that, according to the TV commercial, was supposed to "provide me with a 'colorized' television set. The sheet was tinted with three colors in horizontal bands. The top was blue, the middle a golden brown, and the bottom was green. When I stuck this to the black-and-white screen, I was then supposed to view the pictures in the wonder of color. However, this device took more imagination than I could muster. If the people who marketed this doofy item are still alive, I'll bet they still have a good laugh when they think about how they duped their buyers."

Aficionados knew that special plastic sheets were essential for the full enjoyment of *Winky-Dink and You* (1953–1957), which Alex McNeil's reference volume *Total Television* characterizes as "Television's first interactive show," adding, "The gimmick for which the show is best remembered was the Winky-Dink Kit, which . . . consisted of a piece of plastic that could be placed over a television screen, some crayons, and a cloth to wipe clean the plastic. The kit enabled viewers to participate in Winky's adventures by drawing props as suggested by [host] Jack Barry" and to record clues to an end-of-show secret message.[7] Harold Woodell adds, "At some point during the show Winky or some other character would move his finger in the air in front of the TV camera, and I would trace his movements on the plastic attached to the screen. The result would be a word or a picture that solved a riddle or mystery related to the on-screen shenanigans. My interest in this device waned after the second show." Margie Chapin remembers that "you stuck a green clinging plastic over your screen and used your four special crayons on it to get Winky out of difficulties; much caution from the announcer not to draw on your set without the magic screen or to use your regular crayons." Mary Lynn Moon's mother "would not let us send away for the transparent sheet you put on the screen to trace the clues, so my brother and I would try to keep up on a piece of paper or whatever we could see through on the screen."

"I had a device I ordered for my TV back in the early '50s, something most people don't remember," says Frank Clipp. "It was called a Blab-off. You cut a speaker wire and inserted this thing; it had an in-line switch in it, and you ran it over to your chair. When the commercial came on, you just cut off the sound. It worked!"

Marshall Y. Feimster, Jr., spent the 1950s in the U.S. Air Force, and, because of his military duties and the uncertainty of finding a working TV set, his experience with the medium during that decade was hit-or-miss. Some barracks had TV receivers, but they "usually didn't work." Stationed at middecade in Miami, he says, "we would all gather at the NCO club to watch Phil Silvers as Sgt. Bilko and *The $64,000 Question.* Sent to Nellis Air Force Base for an atomic bomb test, he had his first experience with twenty-four-hour viewing, and in the NCO club at Nebraska's Lincoln AFB the preferred video fare was "Andy Williams as a summer replacement" in 1958. At the end of the decade he "scraped together $125 to buy the first TV, a 13-inch black and white with rabbit ears." In subsequent viewing he especially enjoyed "Sonny and Cher, costumes and all."

June Elsey still has the seven-inch black and white Motorola set that she watched as she grew up on a Pennsylvania farm. "It still lights up, and I have tried to get tubes for it, even called Motorola, but have had no luck," she says. Watching that first television "was a big deal, and we did have neighbors over to watch certain programs."

The arrival of a TV set in Dr. Cindy Roden Goodloe's childhood home forced a major choice:

> As I was entering second grade at Irvine Elementary School in Orange County, California, I made the momentous decision NOT to continue in Brownies that year, 1960. (Please understand that that was a big deal back then because Scouts would actually wear their outfits to school on meeting days, in my case a little brown dress and beanie cap, all festooned with pins and patches for achievement.) So why would I want to quit? Well, we had gotten a brown Zenith TV (it sat on big metal legs), and I was completely enthralled with *Popeye*. (And, yes, I can credit a lifetime of loving spinach to that show.) In any case, the Brownie meetings were going to be AT THE SAME TIME AS *POPEYE*, so there was no way I could continue Scouting. By spring, I was very upset about quitting, as I found out that the Brownie trip that year was to Disneyland.

For many years Cindy Goodloe understood that her father had relented in his firm opposition to television and had bought that Zenith set. In checking with her mother recently, she discovered that "my father never did 'give in' on buying a TV. Instead, our next-door neighbors at El Toro Marine Base felt sorry for me and my sister (not having a TV) and gave us their TV when they moved away. My father (age 82) is still extremely anti-TV and seems to have given us all an anti-TV attitude."

Color television sets became reasonably affordable in the early 1960s, and the competition to be "the first in the neighborhood to have one" seemed to cycle around again, although questions of quality improvement and an inevitable price drop once more nagged potential buyers. Beth Jarrard's family had a divided experience here:

> My dad sold TV sets in his hardware and furniture store. He became one of the first retailers in our part of the county, and we were one of the first families in town to have [a black and white set]. Sadly, we were not the first to have a color set. Dad was convinced the technology would improve if we waited just a little longer. He was right, but the thrill was gone by then. At our hardware store, Dad prided himself on fixing anything he sold, so he employed a man to repair radios and TVs. Mac's workshop was always filled with sets in varying stages of repair, and he made regular calls to our house to improve the picture.

Germaine T. Leverette remembers the great anticipation when her parents purchased their first color set:

It was a huge event, and it was a Christmas present for the whole family. The afternoon it was delivered to our house, my parents hooked the television up according to the directions and turned it on. My brother and I had been promised that we could watch *Rudolph, the Red-Nosed Reindeer* as our first show in color. Everything worked properly except for the sound. My parents ended up playing the black and white TV in another room for its sound while we watched the new TV for its color. I have to say the real color was a large improvement over the plastic cover overlay we used to have on our black and white TV. To this day, every time I see or hear about the classic *Rudolph, the Red-Nosed Reindeer* show, I fondly recall our first experience with color TV and still have to laugh that we had no sound on the new TV.

Roger Rollin recounts an episode in which an inspired flash of teenage irony seemed almost to pull a new TV from the sky. A teaching colleague's son "had been bugging his dad about getting a color set, and the father, of course, like any good parent, saw no reason to spend good money on some fancy new technological advancement when the old technology worked perfectly well. Then one evening father and son were just leaving a local discount store when the father gazed upward and said, rapturously, 'Look at that beautiful sunset!' Replied the son, 'Oh, I don't know, Dad; it would be just as good in black and white.' The father turned around, went back into the store, and purchased a color TV."

In Andy Oliver's family, television (whether in black and white or in color), food, and proximity served to bind generations:

For my grandparents, television was a privilege, a reward for sacrificial services rendered. They had worked hard in the local textile mills in our town all their lives, raised a family of five children during the Depression, and had known few of life's finer things during those lean years. I believe TV was their just compensation.

Some of my earliest memories involve "Popaw," "Momaw," and a house full of cousins warmed by gas logs, fellowship, and a black and white TV. After a day of chores and children's games, my grandmother would prepare supper for us all as Bob Ledford, a used car dealer from Asheville, North Carolina, sponsored and, I believe, hosted "wrasslin'" on Channel 13. This was merely the warm-up for an evening of serious television viewing. Boy Scout troops and church groups lined the ring as late 1950s-early '60s gladiators circled, then engaged their adversaries in what I perceived as a fight to the death. I can still hear enthusiastic Cub Scouts screaming, "Go get 'em, Johnny!" This was Johnny Weaver, of course. His is the only [TV wrestler's] name I can remember.

After supper we settled in for our Saturday evening TV marathon. I'd recline on the floor, toasting my toes before that strange-looking blue and orange gas fire as we watched shows like *Wagon Train, Rawhide*, and for a change of pace *The Joey Bishop Show*, which I vaguely remember as a comedy similar to *Dick Van Dyke*. The evening usually ended with a game of Old Maid, with *Saturday Night at the Movies* as our backdrop. Does life get any better than this?

It did! In 1963 my grandparents built a house next to mine. With the purchase of their new brick home, they also decided to acquire . . . you guessed it—a new TV. But this was not just any television. It was "IN COLOR!" I couldn't wait to see it. The set was delivered late one evening. I'd been unable to concentrate at school that day. My thoughts were consumed with the fact that I would have unlimited access to a new state-of-the-art color TV right next door. (We still had black and white in our home. In fact, it may have been late '60s-early '70s before we took the plunge. Mom and Dad chose feeding and clothing us above entertainment. I appreciate that . . . now.) As I tried to corral my wandering thoughts at school, I can still remember those top questions and concerns: "What will the NBC peacock look like?" and "I wonder what black and white will look like in color." We gathered around the new set that night. I stood amongst parents, siblings, aunts and uncles, and my questions were answered. The peacock was . . . it was . . . well, it was vivid, and black was black and white was white. What a moment! They let me watch *Jonny Quest* that night—a primetime cartoon. The colors were breathtaking.

My dreams and expectations surrounding my grandparents as neighbors were quickly realized. I would enjoy double suppers. Biscuits, mashed potatoes, fried steak and gravy, country ham, sausage, pancakes, and an assortment of fatback-saturated vegetables could be savored in one night by eating at home and next door. My grandparents, of course, pampered me when my parents failed to realize how privileged they were to have me under their roof. Their home was a refuge when my brother and sister threatened me with fists and claws respectively. Finally "Popaw" and "Momaw's" toasty living room was my place of limitless TV viewing pleasure. I knew every setting on my grandparents' rooftop rotary antenna, and every night I was their remote control as I found myself situated on the floor before the multicolored tube watching Westerns. Our favorites were *Gunsmoke* and *The Virginian* with a host of other programs supporting our frontier appetites. This was, I believe, the heyday of the Westerns. You could watch them almost anytime, and on any channel.

My serious viewing days with my grandparents lasted into the medical show era as well. As Westerns dwindled, we simply eased on into *Medical Center* or ventured into the TV lobby of *Marcus Welby, M.D.*

I savored holiday fare with my grandparents also. *Bing Crosby Christmas Specials* were anticipated and scheduled weeks in advance. *Frosty the Snowman* was endured by my grandparents for the sake of the grandkids. *Rudolph, the Red-Nosed Reindeer* was my favorite then and still is 'til this day. I saw it first at "Momaw" and "Popaw's," and I've made an effort to see it every year since. During my college years I once opted to watch Rudolph rather than attend a Clemson-Furman basketball game. My masculinity and maturity were questioned, but my spirit was quite satisfied, and my nostalgic soul was nourished.

Those were some wonderful days and nights with my grandparents, and I couldn't tell you when they concluded. In fact, they didn't end. The moments before the television together just simply tapered off. Those extra meals had placed me in the fast lane on the road to obesity, so I didn't stop in for supper as often as I had in the past. My brother, sister, and I began settling our differences without my grandmother's help, and Mom and Dad finally reached the point where they would purchase a color TV for our family. My social life began to increase, and slowly the evenings with my grandparents began to decrease. Sadly enough, this is the way it's supposed to be, I guess.

Dick Stanley thinks of a 1973 incident that divides the pioneering black-and-white-only viewers of the first TV decades from their successors. His mother was preparing to move into a condominium and invited him to take anything he wished from the attic before she held a yard sale:

> One of the items we took was an old 19-inch black and white portable TV that my brother had used in college. My daughter, six years old at the time, was stationed in our bedroom, where the "new" TV was set up. She was talking to me through an open window while I was on the roof adjusting the antenna. At one point, after turning it back and forth several times and listening to my child tell me "That's better" or "That's worse," I heard her finally say, "That's a pretty good picture, Daddy, but the color is terrible!" The kid had never seen a black and white television.

In fact, that generation was spared even the difficulties of installing and tuning the earliest color sets. In the first half of the 1960s, Ronnie Rahn's father owned a small appliance and TV store in Sumter, South Carolina, and the son enjoyed helping his father with deliveries after school and on the weekends:

> The most fascinating of the deliveries were the color televisions. There was a set-up process that started with the homeowner having to commit to a spot in the house where the TV would be placed. The set had to stay in that place and couldn't be moved because the electronics were extremely sensitive, and moving the set would mean another service call to realign the color "guns."
>
> Once the TV was set in place, the back was removed, and a color bar generator was connected. The TV was turned on, and the technician would switch the generator to make different patterns appear on the screen. As each pattern appeared, he would adjust the red, blue, and green colors to align on top of each other. He would switch from vertical lines to horizontal lines to crosshatches to dots. When the colors were properly aligned, the pattern would be white, and there would be no evidence of one of the colors showing on the edges. This was a tedious process, and moving or bumping the TV would throw the color "guns" out of alignment, so the homeowner had to understand that the TV stayed put. The whole ordeal was something to watch. After the generator was disconnected and the back of the TV was replaced, the next step was to turn on the TV to a color show and adjust the tint and intensity to get the correct hues and tones.
>
> This process took about 30 to 45 minutes if we were lucky. But in those days it sure seemed worth the trouble when the Cartwrights rode up on their horses at the beginning of *Bonanza!* Nowadays we take so much for granted. We go to the store, buy a TV, throw it in the back of the truck, and plug it in when we get home, all without the wait for the deliveryman and the technician.

Notes

1. Myrna McKee, "Meeting Mister Television," *Daily Journal* (Seneca, SC) (December 14, 2000): sec. A.
2. Leah K. Glasheen, "Reconsidering the '50s," *AARP Bulletin* (January 1998): 20.

3. Transcribed by Corinne Holt Sawyer (Carlsbad, CA) (May 15, 2000).

4. Untitled single-panel cartoon, *Collier's*, 134(1) (July 9, 1954): 84.

5. Walter Hill, Foreword to Ric and Stebbo Hill, *TV Photographer, 1957–1962* (Macon, GA: An Obscure Publication and Atlanta: Southeastern Arts Media and Education Project, 1988): 1.

6. Mitch Fields "Incoming: Nowhere to Surf," *New York Times* (February 18, 1999): sec. D.

7. Alex McNeil, *Total Television: The Comprehensive Guide to Programming from 1948 to the Present*, 4th ed. (New York: Penguin, 1996), 918.

$$\text{——}\ 3\ \text{——}$$

Antennas, Rotors, and Hope

In the precable television era, the antenna attached to the roof or clamped to a pole in the backyard was a proclamation: "We have a television set!" City folk might receive acceptable video and audio signals with the thin telescoping shafts of "rabbit ears" extending two or three feet from the back of the set's case, but more distant viewers and TV installers became anxiously acquainted with outdoor antenna shapes and brands, rotors, wood or metal mounting poles, and other paraphernalia that might reduce or eliminate reception snow and the annoying abstract-expressionist patterns (accompanied by chugging and whining interference noises) generated by passing airplanes, trucks, and, it almost seemed, large dogs with metal collars loping through the yard.

The eager business of raising, positioning, and securing the antenna on its support was often the unpaid collaboration of friends, relatives, and neighbors. When that task was well-accomplished, husbands (in that postwar day of conventional *Father Knows Best* and *Ozzie and Harriet* family roles) could take a provider's pride in the antenna's skyward thrust; it got the reception job done. Wives often found it a tacky unbalancing of the house's street-view symmetry, an unwelcome focal point for visitors to notice, and an implicit invitation for TV-less distant friends and relatives to drop in for a visit. The children were simply delighted to have access to *Romper Room* and Hopalong Cassidy movies. For a short time, at least, the kids also had a new game: Twist the direction-setting knob around the clock-like face of the Channel Master control box atop the TV set, run out into the yard, and watch the antenna rotate toward Lubbock or Hartford in slow mechanical majesty.

Some reception strategies were closely calculated; others sprang from long trials or emergency expedience. In Spartanburg, South Carolina, Jewel Mason (according to a recounting by her son-in-law, Warren Edminster) "had received a new bicycle for Christmas and absolutely loved it. A few weeks after Christmas, the antenna to her family's TV blew down, and her father promptly commandeered her bicycle wheel for antenna duty. It worked just fine. She wasn't thrilled about the loss of her new bicycle (restored to riding condition later), but she received some solace from the fact that the family set was working again."

Owens Pomeroy knew many "trials and tribulations" in television reception in Baltimore:

> You would have to adjust the rabbit ears for each channel. If you lived in the inner city, forget about good reception because of the many wires and electric outlets. We were fortunate enough to live in the suburbs and got pretty darn good reception. I guess living near the TV studios and towers helped a little bit; WJC, WBAL, and WMAR were within a mile of my house. The problem with the indoor antennas was solved later with the introduction of the roof antenna ($39.95 complete installed, according to the 1949 newspaper ads, and for $10 extra you could have a rotary added.) Other problems were cars idling, planes flying overhead, power tools being used while the TV was on. All these factors would cause a herringbone interference with your set, and sometimes if a person walked across the floor quite heavily, the picture would jump.

Jack French reports that two years after his first high school days' encounters with snow-filled television pictures,

> the two Milwaukee TV stations boosted their signals sufficiently that programming could be received well enough in our area [Rhinelander, Wisconsin] for many folks to invest in TV sets. Although the knob on our set contained 13 channels, we could get only two . . . and those very poorly. In those days a great deal of attention was spent on arranging the rabbit ears in different directions until the image came in the best. With one person watching the set, another person would slowly move the rabbit ears in different directions until the image came in the best. A strange custom of securing strips of aluminum to the rabbit ears began, under the premise that this would improve reception. Whether it did or not, the custom spread from one household to another, and soon it was rare to visit anyone who did not have their TV antenna decorated with flags of aluminum foil.

The issue of television reception half a century ago often amounted to "location, location, location." Marilyn Chadwick reports that her husband remembers "seeing all these funny-looking wire contraptions appearing en masse on the rooftops in the town of Mexico, Missouri, and, upon inquiring, he quickly learned that these were antennas and was impressed with their function and purpose." Much of Margie Chapin's childhood viewing in the mountains of central Pennsylvania was tethered to a single channel because "that's all our low antenna could pick up," but she did manage to keep current on *Miss Frances's Ding Dong School* and *Captain Kangaroo*. Dixie Ann Schmittou's recollection of her childhood in middle Tennessee illustrates how great a problem reception could be in an area of varied terrain:

> My family had been hooked on radio, so it was only natural that when television became available in 1949 we would do whatever it took to bring it into our home. Believe me, that wasn't easy. My daddy farmed and also worked as a carpenter in Nashville. The NBC radio affiliate there (WSM) was the first station in the area to broadcast TV signals. We lived on a farm in Ashland City, just 15 miles from Nashville. The problem was reception. Our house was on a high hill among other high hills. No matter the antennae, reception was never very good. But at the time, just being able to see it at all was marvelous.
>
> My mother became our TV repairman. She noticed what the repairman did and later on would buy the tubes on her own and replace what was needed. I still regret that she didn't live long enough to have cable and, at long last, a good signal.

As Harold Woodell observes, some families sank most of the TV budget into buying the biggest-screen set that an already strained conscience allowed, so the antenna was often an afterthought issue which, ghost-like, reappeared each time a transmitter in a different direction was to be accessed. While the TV bourgeoisie lounged indoors on a cool evening and watched the electric rotor's control box pointer click off the direction markings around its clock-like face, the head of an "economy plan" TV home, or a delegated child, would have to rush into the cold or rain, put one or both hands on a cold metal pole, and twist according to the advice shouted from a window on the windy side of the house: "A little more to the south, Billie," or "No, no, Channel 11 is THAT way."

Sometime in the 1950s, about ten years after he had first glimpsed a television picture in a Pittsburgh hotel bar, John F. Clark was working in a small town on the Gulf Coast of Texas, about halfway between Galveston and Corpus Christi, and he volunteered to help a neighbor raise an antenna to serve his new TV set. The neighbor "had been told that it must be high enough that there would be no obstruction between it and the antenna of the sending station. There is significant curvature in the Earth's surface in those hundred miles. He bought the longest telephone pole that he could find, and we erected it behind his house. I was at the top of that pole turning the antenna one way or the other while Phil watched the screen to see where there would be the least 'snow.' It was then that a 'Blue Norther' came through, and the temperature dropped from a hundred degrees to about fifty in the time that it took me to climb down that pole. I thought we were going to have some *real* snow."

Even in a midsized flatland city an antenna was often necessary, as Bill Koon's experience in Columbia, South Carolina, illustrates. Tired of visiting relatives to see the tube, he says,

> eventually we got our TV, a Meck which we bought at a furniture store, which was appropriate because it was far more furniture than TV. We put up a roof antenna and were constantly out there turning it to try to get the signal when programming started

at 5 P.M. We were briefly the hit of the neighborhood; everyone came to our house where the best entertainment was probably watching my father try to turn the antenna and/or adjust the many knobs on the back of the engine so that we might get some faint picture behind all the snow. Long after the device established itself and we could all watch all the TV we liked, I became an intellectual and refused to allow it in the house. But I got over being an intellectual and now watch TV as faithfully and as mindlessly as anyone else.

Married veterans taking advantage of the GI Bill's education benefits in the late 1940s and early 1950s often found that the best housing option was to move their young families into "prefabs." These small wooden structures had been designed as housing for bombed-out people abroad, but they were later offered to colleges and universities as instant living spaces for the influx of older students. Jim Hattaway recalls his situation at Clemson in 1952, when his wife worked in the school's Housing Office and they lived in the "prefab" fringe of the campus:

> At the start of the fall semester, my friend Leo Lindell returned from home—Brooklyn, N.Y.—with a used TV set, a fork antenna, and a Model T Ford (another story.) Well, Leo couldn't use the TV in the Barracks, so he unloaded it onto us at the "prefab." Leo knew he was always welcome and usually showed up at dinnertime!
>
> We rigged the antenna on a 20-foot bamboo pole and nailed it to the side of the "prefab" using old license plates! We set the TV up in the spare bedroom. For days we tweaked and adjusted everything and anything a screwdriver could reach. One Saturday night my wife said, "Give it UP! You're wasting your time!" and went to bed. Leo kept adjusting, and all of a sudden we had SOUND—and a picture! It was "Snooky" Lanson on the *Hit Parade*, singing "How Much Is That Doggie in the Window?" (Real Quality TV.) I dashed in and woke my wife, and when I got her in there, only snow was showing on the set. Leo had over-adjusted something! From then on, we enjoyed only radio at Clemson!

Even several student generations later, leaving home for college brought the location issue home to Judy Davies, who says, "As a child I grew up in the New York City area, and even before cable we had lots of channels. I thought everyone all over the country had lots of channels too and learned differently when I went off to Clemson and we couldn't even get one channel in our dorm rooms (1975). I learned fast that people in areas outside of major cities did not watch TV very often because of this. Also, I remember visiting my grandparents as a kid (1960s), and although they had a lot of channels too, the reception was so poor that they had purchased a remote control antenna that made an incredibly loud noise when you were trying to bring in a station!"

No sooner did U.S. viewers become adept at aiming their antennas toward high-powered Very High Frequency stations' transmitters than the advent of smaller outlets in the Ultra High Frequency range (received on channels numbered above thirteen) necessitated a retrofitting of reception devices. So it was that Walter P. Horlyck, "back in the mid-'50s when the world was more innocent and everything was either black or white," simply wanted his television set in San Francisco to pull in a watchable picture for broadcast baseball games. He continues,

> Yes, it was also the time of McCarthyism, the Cold War, and so on. I was an appliance repairman for Sears, [which] had gone into introducing UHF in their Silvertone TVs. In order to receive UHF, it was necessary to have a UHF antenna installed, in addition to the VHF one. The store I worked in was going into a big promotion of UHF sets and antennas. Bad feature, though: there were many critical areas for reception. A UHF signal goes in a straight line, and the San Francisco area, with its weather and hills, created some problems. The store was installing antennas for employees living in those critical areas. Since we lived in an area that was really borderline, we got one on a trial basis.
>
> As it happened, I was a rabid baseball fan, and the San Francisco Seals (before the Giants) were televising their away games on a rinky UHF station out of Oakland, across the Bay. With the weather, the reception would change. Since the TV was in the front room adjacent to the fireplace, I would go up on the roof and adjust the antenna while shouting down the chimney to my wife until I got the picture perfect. So happened we didn't have the best of relations with one of our neighbors, and one of our friends across the street came over one weekend and in the course of the conversation said So and So was saying that he had it on good authority that I was sending signals and receiving them from the Russians on that strange antenna-like thing I had on the roof. I explained what it was and posthaste made it known to the whole neighborhood before old Joe McCarthy could come from Washington, DC, to have me investigated.
>
> That antenna was the bane of my existence.

Randy Cox offers a summary view of a generation's leap from antenna reception to the cable connection:

> I don't usually think of myself as being that old. Old people are folks who grew up with radio, or even less. They were children who had to stoke the fire each morning just to have the means of cooking on the woodstove. At least that's the way I remember it because that's how it was for my grandparents, and to some extent, for my parents. So, I can't be old.
>
> But when it comes to modern technology, I guess I'm a geezer. Current generations grew up never knowing life without a microwave or a telephone answering machine. "Ghosts" and "snow" are terms in no way related to their remote-controlled televisions, and they have never asked Mom to turn off the electric mixer because it was interfering with the picture. Hell, they've probably not seen an electric mixer, because their box of microwave brownies comes premixed; "just add water."
>
> I guess my distant memories of watching three-station television are part of a dying social collective consciousness. When cable television appeared, I was already a 22-year veteran of watching broadcast images which traveled over clotheslines to rooftop antennae, both of which no longer dot our landscapes. When we progressed to the world of the rotary antenna, we would wait a few minutes for it to reposition

from a northwestern orientation to a northeastern one, as you couldn't progress in either direction across the northern orientation [the 12 o'clock position on the rotor mechanism outside and the directing dial beside the TV set]. Today, with DirecTV and 500 instant channels, I doubt that this delay, just to check out what was on another station, would be tolerated. Instant gratification wasn't a hallmark of 1960s and early '70s TV.

— 4 —

TV Behaviors and Protocols

The installers had put the new television set into place and tuned it to the station's test pattern. The outdoor antenna had been secured by guy wires or has been judged sufficiently well braced to withstand any wind. Could the pioneer TV-owning family circa 1950 now enjoy its programs in relaxed privacy? No. Not likely.

There were family issues of when, where, what, and how much to watch. There were curious neighbors. There were relatives. There were female guests who came in Sunday outfits and gentlemen who arrived in suits and neckties merely to have the pleasure of being introduced to the video marvel. Children, some from tribes entirely unknown to anyone in the household, swarmed like gnats. Some older guests arrived by invitation, others getting a foot inside the door with a hasty "Oh, I just thought I'd pop in for a moment. What's on?"

In southeastern Atlanta, a meticulously groomed white-haired neighbor, Mrs. Elmer F. Loveless, joined the L. D. Strom family to pose for Hugh Stovall's newspaper photograph of viewers gazing at the new console TV in only slightly relaxed dignity. "Be hospitable and share your TV set with someone who doesn't have one" seems to be the theme illustrated in the intense gazes of father, daughter, mother, neighbor, and, at their feet, Taffy, the raptly attentive cocker spaniel. When the picture appeared in an *Atlanta Journal-Constitution* special Sunday section in the late summer of 1951, it was followed by short paragraphs of bright prose celebrating the happy expected future of home (and neighborly) TV viewing.[1]

When television came to Greenville, South Carolina, Wilma Edmonds and her husband were great enthusiasts who "would go down to Augusta Road and watch TV in an appliance store window" or ride over to the WFBC-TV studios and watch the monitors there. "Our real excitement came," she says, "when we were able to purchase our own 'BIG box with its little screen'" and to have their first TV guest: "We lived in a two-room apartment, with our landlady [living] in the other side of the house. While my husband and I had already been enjoying our TV and staying up late to watch whatever closed out the evening—we even looked at the test pattern for a while—our real fun came when we invited our landlady to come over and watch. Wow! Did she ever enjoy our little treasure. We sat in the kitchen, in straight-back chairs, watching wrasslin' and cowboy movies until the proverbial test pattern came on. One would only have had to be there to see and enjoy her reactions to [wrestler] Gorgeous George as they faked us out—as they still do. She fell out of the chair one night."

Some guests apparently cared more about TV-viewing than about visiting with their hosts, as Mrs. C. H. Blocker, Sr., discovered while her husband was a grocery store manager in Charleston, South Carolina, from 1951 to 1953. Before the first local station came on the air, he won a twenty-one-inch console as a prize for highest sales in his area. The only available channel was located in Jacksonville, Florida, and reception from that point 196 miles down the Atlantic coastline was quite variable. "We had a friend, a Navy chief friend," she says, who would frequently call to ask if the reception was good. "Well, if not, he 'won't come tonight,'" she reports. "He would only come when the reception was good."

In Mary Ann McKenzie's "middle class polyglot neighborhood" in White Plains, New York, TV sets were scarce, and at the behest of her younger brother, a Roy Rogers fan, "lots of kids that I didn't know showed up to watch" the after-school programs every weekday. "My mother says she didn't know them either," she continues, "but she was stuck with them when her own kids went out to do other things."

"We wore out our welcome at the Kellys'" in the early 1950s, Howard Porter says of himself, his older brother Charles, and his younger brother Roy. He explains:

> Our nearest neighbor was Lake Kelly, and my brothers and I were fascinated when he erected a huge antenna above his house and bought a TV set. We never missed a chance to get invited to stay and watch Uncle Miltie or Red Skelton on their 12-inch screen. My brothers and I begged my dad to get us a TV, but the two stations Mr. Kelly could get on his huge antenna were in Atlanta and Charlotte and were very snowy. My dad told us that he would get us one when you could see the whites of their eyes; he was sure, of course, that would never happen. After about six months of Mr. Kelly having to run us home almost every night, Greenville and Anderson stations went on the air, and we practically dragged my dad to Mr. Kelly's house so he could see the whites of their eyes. True to his word, when we came home from school the next day, our dad was on the roof attaching a huge antenna to our chimney, and inside the house sat a brand new TV. I always suspected that Mr. Kelly had a lot to do with it so he would not have to run us home every night.

In Dixie Ann Schmittou's childhood home, late-staying TV watchers were an inconvenience. "My dad's boyhood friend and his wife used to come over almost every night to watch the news," she says, "and Daddy wasn't too happy about that. He had to get up early the next morning, build the fires, milk the cows, and then be off to Nashville for his job. I can see him walking into the living room where we were all gathered and announcing, 'You can watch all you want, but some people have to go to work for a living.'"

Garnette Bane remembers that the arrival of television prompted her mother to reorganize her day: "We owned a console TV with doors in the late '40s. Mother would complete her housework before noon so she could watch *Queen for a Day* and other old-timers." Such devoted daytime viewers often would not answer the telephone while favorite shows were on the air.

The middle Tennessee family of Ann Bowers had a very early TV set, and in childhood she fretted a bit as, lifting her body on tiptoes, she passed behind the crowded row of relatives leaning forward in their chairs. Each head blocked the view of others as necks tilted and eyes strained to see the small picture tube centered in the "enormous" cabinet. Her concern was not so much that she could only glimpse the TV screen in gaps between the older and larger heads and bodies. Rather, "I remember thinking even then," Ann Bowers says, "'this will be the end of family life. This is the end of conversation and interaction.'"

Elisa Sparks notes that her mother, while long settled in Seattle, had been reared in Texas and consequently brought a genteel "Southern view" to the issue of placing the first television set in her home. "For my mother," Elisa Sparks explains, "the living room is a place to entertain and converse with guests." Thus the first TV set was banished to the basement, and when the family moved to other houses, the receiver was located in the family room, where it could be seen from the kitchen. "You have to remember that my mother has had all-white furniture in the living room since 1969," Elisa Sparks says, "so her living room was 'not a place to live'" in the casual sense. "It was a class thing: She thought that only lower-class people had a TV set in the living room."

Indeed, the new television set posed new etiquette issues, the primary question being whether or not to turn off the set before opening the door to guests who arrived for purposes other than TV-watching. Mike Newton remembers his family's solution: "In the first house we ever owned (by now it was 1954–1955, and I was 11 and had my own room), the television was kept in a special cabinet where the door could be closed to block it from view when it wasn't on. My mother felt that it monopolized polite conversation, and she never wanted it on when we had guests."

In a newspaper column on her family's adherence to the tenets of Southern hospitality, Rheta Grimsley Johnson recalls a more emphatic deference to guests in Henderson Swamp, Louisiana: "The first rule of the Grimsley household was to turn off the TV whenever visitors arrived. It didn't matter if the Fugitive was about to find the one-armed man, or if Hoss' latest girlfriend was dying on the eve of another *Bonanza* wedding, or even if the relentlessly chipper Mouseketeers had lined up to sing the farewell song. If you didn't jump up and punch the 'off' button, somebody might punch you. 'Turn that thing off,' Daddy would growl, as if we didn't know the drill."[2]

Within many new TV families, mealtimes brought the question of whether to turn off the program and move to the traditional dining area (the better for the quiet digestion of food and discussion of family matters) or to yield to the tube's magnetic pull and set up ad hoc means of holding plates, glasses or cups, and silverware near the screen. The advent of Swanson's TV dinners and the treacherously thin-legged TV tray pressed the issue. This is Marilyn Chadwick's observation: "Indeed, a new era had begun. For the next decade, every newlywed couple received multiple sets of TV trays that folded flat and had crisscrossed legs. They weren't very sturdy, and many a spill was the result of getting tangled up in them with one's feet. In our house they were outlawed for food but permitted for a game of solitaire."

Elizabeth Blakely's parents held to the location line if not the dietary one: "Whenever Mom and Dad went out and we had a babysitter, we were served TV dinners (fish sticks, potatoes, corn), but *NEVER* in front of the TV!" Mike Newton's mother made a different concession: "The first television set my family had was on a trial basis since my father could not afford one. My mother let me eat dinner in front of the set on the first night. This was indeed a special treat as we never ate dinner in front of the set regardless of what the early TV ads showed." Sarah Barnhill and her brother were allowed to have their TV trays near the set until their parents discovered the rotting green beans and other vegetables stuffed under the newspaper lining of the trash basket. "After that," she says, "everyone was seated at the dining room table."

Juana Green grew up in "one of those very small post-Korean War houses" in California. "I would say the living room was certainly no more than twelve feet by, say, eighteen feet, and then right next to that was the dining room, but we never watched television while we were eating. Only if my parents went out for the night and we had a babysitter could we sit there in front of the television with the babysitter and eat, you know, pizza or something, peanut butter and jelly sandwiches, hot dogs. . . . " Today her parents live in a small apartment with adjacent but discrete living and dining room areas, and when she visits, Juana Green shows that she well absorbed the childhood precept: "I have to insist that we shut off the television when we eat."

For the first viewing generation the television set compounded already difficult issues of proper bedtime. "When I was in elementary school," says university development official Jean Kopczyk, "8 P.M. was bedtime—no exceptions on school nights! I can recall having to go to bed just as the *Dragnet* theme music was playing." Mary Ann McKenzie's father, returning to White Plains about 7 P.M. from his train and subway commute to work in New York City, settled down to view his sports programs, and, she recalls, "That always piqued my interest, and I would sneak out of bed and stick my head through the bars of the stairs and watch."

"Most of my early recollections of evening television are based on sound more than pictures," says a high school art teacher whose family in Clifton, New Jersey, across the Hudson from Manhattan, acquired a set in 1952. He explains:

Our house was a small "two family," and we lived on the second floor. My bedroom was about 60 feet from the "sun parlor" where the TV was. My bedtime was 7:30 all through elementary school, and I would be lulled to sleep by sounds like the droning dialogue from Edward R. Murrow interviews or the canned laughter of *Our Miss Brooks, Duffy's Tavern*, or *Corliss Archer*. I also recall the live audience reactions of *The Texaco Star Theater* with Milton Berle.

On the eves of school holidays and on Friday and Saturday nights I was allowed to stay up until 9:00, and if I begged and pleaded, once in a while the limit was stretched to 10:00. Then I was treated to shows like *The Life of Riley* with William Bendix, *The Jackie Gleason Show, Alfred Hitchcock Presents, I Love Lucy, December Bride*, and the list goes on and on and on and on.

"Can a personal TV get *too* personal?" the large type in a 1965 magazine ad for Matsushita Electric's nine- and twelve-inch Panasonic portables asked rhetorically, the supporting text suggesting that the small screens on some competitors' models required strained viewing "at eyeball length."[3] For young Jerry Kaplan, who in the boomtime 1970s and 1980s would become founder-CEO of Silicon Valley's GO Corporation, the question had a different thrust: Does modesty forbid changing clothes within sight of the TV set? In a 1995 book detailing his company's evolution, he recalls a childhood question of TV protocol that eventually grew toward larger issues about the human perception of electronic-screen media messages:

When I was five, my parents got a portable television set for my bedroom. At first I was afraid to get dressed in front of it, for fear that the people on the screen could see me, just as I could see them. I knew that the gray, blurry pictures were merely signals magically sent through the air, but I didn't understand which qualities of living human beings carried over to the new machine. Like the rest of my generation, I was captivated by this limitless window on the world, and what I saw seemed as real to me as my own two hands. Through the power of television, I knew that lions were large, proud creatures that roared and lived in Africa. But one day my father took me to see the big cats at the Bronx Zoo. I could barely relate the coarse, smelly creatures in front of me—lying docile and panting in their cages while flies buzzed around their heads—to what I had witnessed from the comfort of my bedroom. The TV images just didn't capture the real experience.

Sitting in the briefing rooms of AT&T thirty-five years later, watching staffers project spreadsheets, graphs, and slides from their portable computers onto the large screen monitors, I realized that computers mislead managers just as television misleads kids.[4]

Randy Cox, by avocation "a student of TV, a follower of TV, a TV groupie for a long, long time," reflects usefully on the viewing rituals and the status issues of his boyhood home:

Mind you, I didn't know that we were lower-middle-class. I figured we had all the advantages everyone else had, apart from the super rich. But I've learned from others that the world wasn't confined to my hometown. In more exotic places like Sacramento and New Barunfels and Sumter, people in my generation grew up sitting down to dinner with the television set turned off, a practice I've gravitated towards during the child-rearing years. We had TV trays (though we always had a cooked meal for "supper") and actually used them from time to time in front of the set, though that practice was short-lived, due to the cumbersome nature of those flimsy metal devices. Usually we just rushed through the evening meal proper and walked a few feet into the den (family rooms were for larger homes) to watch TV. The house, like many, was even designed so that the cook could watch TV through the opening above the stove.

I actually have adult friends for whom television viewing was a treat, to be parceled out in tiny bits. Other families would make an appointment to see *Mission: Impossible* or *Star Trek* or a specific PBS broadcast. In my neighborhood, this would have been the height of snootiness. Oddly, in my case, a decent amount of money was put into purchasing the set, as it would be used for hours upon hours of entertainment, whereas the people I've met who grew up in the upper-middle-class used second-hand sets or small 19-inch units, because life was centered around discussion and activities, not around the boob tube. This was something I never realized growing up. It took adulthood to discover the other world.

Some families banned the TV set entirely, while others held out for extended periods or established strict viewing limits. Lily-Roland Hall says, "My father wouldn't have a TV until well after I went off to college. We were a reading family. We had to explain to people why ours was the only house on the block without an antenna on the roof." Harriet Hawkins thought that her children "should be outside playing in the sunshine" and did not allow television into her home until her children were well along in grammar school. Her husband was recovering from a heart attack then, and his mother offered a set to entertain the patient. "You know which ones it entertained," Mrs. Hawkins adds, but she discovered that the Thomasville-Tallahassee CBS station's *Romper Room* program, on which local children appeared as guests of "Miss Somebody," "was harmless enough" and lasted only half an hour. Still, she thought 1930s and 1940s radio to be a superior medium to generally "vulgar" television. During his grammar school years Tim Anderson, whose Methodist parsonage family did not buy a TV until 1955, discovered that the black and white DuMont came with restrictions: "I wasn't allowed to turn it on until my mother said so. My preacher dad had a thing about TV on Sunday. I was a teenager before I watched any Sunday TV." His and his friends' parents had a universal command—"Go play"—and in the 1950s, before the advent of video games, the "play station" was a tree to climb or the seat of a bicycle to ride around town.[5]

Ruth A. Watkins managed some early viewing at her maternal aunt's home or when she babysat the next-door neighbors' children, but her own family "didn't get a television set until I went away to school. I came home one summer to find that they had a black and white TV. My father was very strict, and we weren't allowed to watch a 'bunch of junk' [such as] *Peyton Place, The Edge of Night*, and *The Secret Storm*. If a program showed too much skin, was too provocative, or embraced a lifestyle that he thought inappropriate, it was off-limits. *I Love Lucy, Andy Griffith*, and *Lawrence Welk* were among those programs allowed and held fondly in memory. Life absolutely could not go on without watching *I Love Lucy.* I think that was just about everyone's favorite comedy then."

Growing up in a large family on a middle-Wisconsin farm, a promising creative writer encountered a starker discipline: "In our house, TV watching was a whipping offense. There it sat, a black and white TV, the old kind with the big tubes, propped in the living room on its very own cabinet. But, with rare exceptions, we kids were not allowed to turn it on. Exceptions included prior approval for Disney's Sunday night movies or *David and Goliath*, a pretty plotless Sunday morning religious cartoon, or just about anything on PBS." The police drama *Hill Street Blues* was approved for "a brief period" until, "as luck would have it, our folks came in one night at the episode's final scene, which nearly always showed a goodnight kiss in bed between the captain and his lady, and that was the end of our seeing that show too." She, her two older sisters, and six older brothers plotted ways to watch "that forbidden television" after they had completed their own chores and before their parents returned from the barns. They kept the volume low; they left room lights on "so the picture on the screen didn't reflect on the windows or out onto the moonlit snow"; each was prepared to rush to another place in the house and simulate absorption in other activities if the parents' footsteps were heard. However, the elders had growing suspicions about surreptitious TV watching, and they would "sneak up early, with some excuse of a forgotten phone message" or would not extinguish the barn lights, the usual sign that they were returning to the house. Caught, the children were marched upstairs and told to face the wall "while our father came down the line with his belt." The blows fell: "Whop whop whop whop." If the older brothers defiantly mocked their father's punishing blows, he repeated the whippings for all, striking harder. She tried to make sense of the situation: "It seemed really stupid to me, to have a television plugged in and taunting in the corner of the most lived-in room and not be allowed to watch. And I knew then how Eve had felt."

Attitudes toward television surface even today in obituaries. When Leona McWilliams died at age 105 in 2001, a reporter-written obituary article noted that this early-widowed sharecropper managed to rear nine children and buy her own farm and "she took more pleasure in reading her Bible than from any form of entertainment. As for television, 'She said it was the devil's work, too much junk on it,' said her daughter."[6] On the other hand, Oxevene Richardson's obituary described her appearance on *The Price Is Right*—during which she "crowed like a chicken" in imitation of the colorful wild birds inhabiting her Georgia home town—as a "highlight of her life."[7]

Although Linda Howe's family had early embraced TV in the days when "The test pattern was on more than anything else," one anticipated viewing experience nearly precipitated a household viewing crisis: "My favorite moment was the night Ed Sullivan debuted Elvis Presley. My older sister was in high school and thought it was so exciting Elvis was going to be on TV. When he came on [with sideburns and swiveling hips], I remember my father's eyes getting larger than I can ever remember them . . . and a look mixed of shock, horror, and total disbelief. I'm not sure I had even heard before some of the words he uttered that night! That was the first 'garbage' Daddy saw on television. The magic was over."

Notes

1. Martha Smith, "Magic Screen of Television Boon to Family Life in Atlanta," *Atlanta Journal-Constitution* (September 23, 1951): sec. F.

2. Rheta Grimsley Johnson, "Cajuns Are the Most Hospitable of Southerners," *Atlanta Journal-Constitution* (December 2, 1998): sec. C.

3. Panasonic advertisement, reprinted in Jim Heimann, ed., *All-American Ads of the '60s* (Koln, Germany: Taschen, 2000), 399.

4. Jerry Kaplan, *Startup: A Silicon Valley Adventure* (Boston, MA: Houghton Mifflin, 1995), 268.

5. Tim Anderson, "Go Play," *Herald-Leader* (Fitzgerald, GA) (October 4, 2006): sec. B.

6. Kay Powell, "Leona McWilliams, 105, Widowed Farmer," *Atlanta Journal-Constitution* (October 12, 2001): sec. D.

7. "Oxevene Richardson Rites Held," *Herald-Leader* (April 11, 2007): sec. A.

— II —

What They Watched

$$— 5 —$$

An Electronic Vaudeville: Variety and Drama Programs

Early in 1931 Chicago's station W9XAP began telecasting weekly performances featuring vaudeville artists appearing at the RKO Palace Theater. The February 26, 1931, edition of the *Universal Newspaper Newsreel* offered movie audiences a segment ("First Vaudeville Show Broadcast in Television Trials") showing the first of those telecasts being produced (or restaged, as some have argued). In an Internet film loop edited from that newsreel, a magician's right hand draws objects from a felt hat held in his left, and intercut with the magic act are glimpses of the television engineers, whose hands turn large black knobs set in black panels, calling sounds and images out of their paraphernalia.[1] The film editor thus suggests a metaphor: Television is the magic act of the new electronic entertainment world, and its programming variety would mimic the contrasts built into the daily vaudeville bill of fare, where the singing sisters' trio was followed by the ethnic comedian and his dog (Mr. Yip and Mr. Yap), followed by the dramatic sketch or a scene from a full-length play, followed by the showy violinist or accordionist, followed by the magician and his "ever-lovely assistant."

In the late 1920s and the 1930s, the talkies and radio devoured vaudeville, but at the midpoint of the twentieth century, television began to take the shape of an electronic vaudeville, filling airtime with a variety of programs meant to appeal to various levels of comprehension and taste. Scanning the pages of an early *TV Guide* seems to show that the Keith or the Orpheum circuit had been reconstituted in the circuitry of television. Milton Berle, "Mr. Television" himself, was scarcely alone in being a refugee from vaudeville.

Variety, Stand-Up Comedy, and Music Shows

Sorting through his recollections of the medium, Daniel Galvin, a former high school administrator and later a college teacher, declares,

Unique among early programs were the number of variety shows. Ed Sullivan's was the longest-lasting and the best known, cramming as many as twenty different acts in a single hour—as disparate a group as one could imagine: opera singers, impressionists, magicians.

Early television was the domain of the comics. Sid Caesar and Imogene Coca's *Your Show of Shows* was a classic, especially through the use of pantomime. Many of the influential writers of today (e.g., Neil Simon, Woody Allen, Mel Brooks) had their start in this show and shows like it. Has anyone ever offered as droll a look as Jack Benny, answering a crook's demand—"Your money or your life"—with a long pause and then "I'm thinking it over"? Similarly, Benny's tortured semi-playing of the violin elicited the empathy of the common man who was not quite as professional as he/she would like to be in some area of expertise. Red Skelton personified the classic slapstick clown. Jackie Gleason in the "Honeymooners" segments played the frustrated everyman—in this instance, a contentious bus driver, always upstaged by his wife and his sidekick Norton (Art Carney). The closest thing to violence in any of the shows like these was Gleason's threat to send wife Alice "to the moon" [via his clenched fist].

Many of these performers were multitalented. Gleason directed and wrote music. Steve Allen (around for a long time, decrying the violence of later television fare) was a comedian, an excellent pianist, and the first host of the *Tonight* show. On the other hand, for some comics a particular style or line endured, as when Henny Youngman (a perennial guest everywhere) said, "Take my wife . . . please!" George Burns, reacting to one of his wife's frequent malapropisms, got maximum appeal from smoking a cigar and intoning, "Gracie." The humor finally was universally clean and family oriented.

While Dan Galvin's recollections are sweeping, Robert Cox's are particular: "Sid Caesar, who was on live also, kept me in stitches on Saturday nights. I was watching the night he explained to the viewers that his costar had been knocked out cold backstage. Something had fallen on her head." Richard Leiby remembers another hazard of live TV: "Dinah Shore in a big finish, falling backwards over a garden bench, hoop skirt, pantaloons, and all." On the other hand, Richard Smith notes that Milton Berle's program opening looked like a *calculated* disaster: Uncle Milty "used to open by arriving in drag. The Carmen Miranda motif (bananas on the head, modeled after the headgear of the female star in the Bob Hope-Bing Crosby 'road' pictures) was perhaps his most outlandish costume."

"The variety was astounding," Marilyn Chadwick says of her early TV recollections and especially of CBS-TV's Sunday evening fixture *Toast of the Town*, with newspaper columnist Ed Sullivan as host. That program, she continues,

was the highlight of the week in most televisioned households. Mom did her homework well in advance, so as to know who the upcoming guests were; and we were generally privy to this disclosure while eating our evening meal. The set was not turned on, however, until the table was cleared and the dishes were done. We took our usual seats in the living room, Dad in the brown chair, Mom in the rose-colored chair, me on the sofa, and Diane on the floor in front of the TV.

There were acrobats, jugglers, trapeze artists, ballerinas, tap dancers, sword swallowers, knife throwers, and unique characters like Victor Borge, who was a comic and a pianist. We were entertained by Red Skelton, Pinky Lee, Norm Crosby, Milton Berle, Alan King, Flip Wilson, and Phyllis Diller. We were serenaded by Nat King Cole, Dean Martin, Robert Goulet, Steve Lawrence and Eydie Gorme, Vicki Carr, Jack Jones, and Kate Smith. (Mom read that Kate Smith had a fear of being approached in an unwelcome manner and carried a pistol on her person at all times. I remember trying to figure out where it could be hidden. Being a seamstress, I envisioned secret pockets sewn into the folds of her gown.)

Teresa Brewer was always announced as the little girl with the big voice. Then there was Jimmy Durante. He had the gravel voice and the big, big nose. He could sing and dance and had a style all his own. I can still see him tipping his hat as he made a comment like "It's just a FIG NEWTON of my imagination," a saying, incidentally, which my family members still use today in honor of his memory.

My first exposure to Elvis Presley was on Ed Sullivan's show as well. Ed was visibly uncomfortable with the music as well as the moves. It was well-known that Ed ran a pretty tight ship and was not enamored with the trends he saw taking shape.

While the Sullivan program was a part of the Sunday evening ritual for Marilyn Chadwick's family, *The Jackie Gleason Show* was the must-see Saturday night entertainment:

On a one-to-ten scale, this was a solid ten! Jackie played so many characters: the Poor Soul, Reggie Van Gleason III, Joe the Bartender, and last but far from least, Ralph Kramden in the Honeymooners sketches. I'll never forget the episode with Ralph trying to teach Norton how to play golf. He told him that before he took a swing, he should first "address the ball." Norton saluted and said, "Hello, ball," and Jackie lost all patience with him. For years since then, we continue to address the ball Norton-style whenever we play miniature golf.

Jackie was a big man, but very light on his feet. One night as he was doing a little dance step off the stage, he fell down. We laughed, thinking it was a planned part of his performance, but we later found out that he actually fell by accident and broke his leg.

The dancers that stole the spotlight on his show were called the June Taylor Dancers. They were my inspiration for taking tap dancing lessons. Our instructor patterned some of our routines after their particular style.

Some viewers were unprepared for the range of variety show guests. A Southern observer recalls his 1950s weekly viewing opportunity with mixed feelings:

We didn't have a television set when I was growing up, and the only program I watched as a kid was *The Ed Sullivan Show* on Sunday nights. My mother, father, two brothers, sister, and I would pile into the pickup truck, go to my uncle's house, and crowd around the black and white set. For the most part we simply enjoyed the program, but occasionally a black performer would be on the show (for example, Sammy Davis, Jr.), and Ed Sullivan would shake his hand. My uncle, who was one of the worst rednecks I have ever known, would get so angry he would turn off the set and that would be it for the evening. He couldn't stand the sight of a white man touching a black man (shades of Archie Bunker, but for real.)

"Of course I remember the variety shows very well—and my folks loved them too," says Tom Worsdale, who settled in San Antonio after a twenty-four-year Air Force career and now runs a music business with his wife Maggie, a professional singer. He found personal inspiration in the shows that he watched during his Brooklyn youth:

Our favorites were Dean Martin (Thursday nights) and *The Jackie Gleason Show* (Saturday nights). Everyone in the country watched these guys along with Ed Sullivan and Red Skelton. I think these variety shows had more long-term impact on me than anything else. I eventually went into theatre and music and to this day have a great appreciation and knowledge of the old standards and Broadway music. And I love the old short vaudeville sketches that were so prevalent on those shows. Who can forget Jackie Gleason doing the Poor Soul or Joe the Bartender with his sidekick Frank Fontaine as Crazy Guggenheim? Or Dean Martin playing the always slightly crocked singer with the Golddiggers? And the guests—always the top singers, comics, and even contemporary rock bands. This is an art form we have lost today, and it is too bad, but I think it may come back, and I bet it will be successful.

Many variety and comedy shows had begun on radio and in the 1950s were adapted to television. Art Donahoe counts *Blondie, Fibber McGee and Molly*, and *Duffy's Tavern* as his favorites among these, noting that he still uses the expressions "Perish forbid" and "Heavens to Murgatroyd" from the last named show. Richard Leiby observes, "Jack Benny's trips to his vault were much funnier on radio than on TV." *Amos 'n' Andy*, having originated in the blackface traditions of vaudeville and amateur theatricals, was voiced on radio by Freeman Gosden and Charles Correll, who had begun their careers in the 1920s as traveling organizers of local talent shows. For the 1951–1953 television series it was given an African-American cast and amounted to "harmless racism" in the view of James Andreas, Sr., a longtime student of ethnic diversity. (A decade later, however, CBS withdrew the program from syndication under mounting pressure from the civil rights movement.)

Your Hit Parade debuted on radio in 1935, and far-flung World War II military personnel valued it as a way of keeping up with popular culture at home. An eight-year NBC television stint began in 1950, with brief revival attempts on CBS in 1958 and 1974. "As a teenager in the mid-1950s," Robert Cox says, "I felt as if I were really growing up because I liked to watch Dorothy Collins, Snooky Lanson, Gisele MacKenzie, and Russell Arms belt out the top seven pop tunes of the week on *Your Hit Parade*." On a trip with her parents to New York City, Linda Law saw

Gisele MacKenzie and Snooky Lanson in the Rockefeller Center cafeteria and got their autographs: "Wow, was that hot stuff!" Albert Holt was impressed with the program's production values. He recalls the elaborate set evoking Fifth Avenue "with 1920s costumes and parasols" for Irving Berlin's song "Easter Parade."

"We never missed *Arthur Godfrey's Talent Scouts* or *Ted Mack's Amateur Hour*, the latter being raw talent of every sort," Marilyn Chadwick says. "Every week it was something new to look forward to, and I very much liked the personality of Mr. Mack himself. There were no frills. The acts appeared on stage, did their dance, sang their song—no wild lighting effects, no tricky camera moves. I preferred it that way. My most memorable performance was a little girl's rendition of 'Pennies from Heaven.' Boy, could she ever belt out a tune!"

The crooning style developed by Rudy Vallee, Bing Crosby, Ray Eberle, and others in the 1930s was sustained in the early network television era by Nat King Cole, the genial Perry Como, Dinah Shore, and other headliners whose ballads alternated with "cute" upbeat novelty numbers. Frank Sinatra's two early 1950s series were not especially successful, but his later specials and guest appearances were much anticipated. At a very different point in the popular music spectrum, older viewers relished the Eastern European-via-U.S. Midwestern "champagne music" of Lawrence Welk's band in a rigidly wholesome presentation of polkas, sweetly harmonized hymns, and perkily staged novelty numbers. To project an image that would later be called "family values," Welk was known to exercise strict control over his performers, and the romantic ballads sung by even the most buxom members of his cast had a curiously unerotic clean-scrubbed quality. "One of my favorite recollections of my early TV viewing was in 1961 when *The Lawrence Welk Show* was popular," Joyce Bridges says. "I was ten years old, and every Saturday night when the show came on my entire family would sit around the TV. It was a time for togetherness that was so special to me. To this day when I see a rerun of the Lawrence Welk program I get that same feeling I had back 40 years ago." Beth Jarrard identified with the star of *The Kate Smith Show:* "I was always a big girl. Kate was a big girl too, but she was also a big star, and I loved hearing her belt out 'God Bless America.' This show also had a princess, but instead of the Indian princess of the *Howdy Doody* show, Kate Smith's Storytime Princess was like a fairy princess dressed in beautiful dresses, and she wore a crown. I liked that part."

Like Lawrence Welk, CBS's Arthur Godfrey was known to be highly exacting with the cast members of *Arthur Godfrey and His Friends*, and among these was crooner Julius LaRosa, who first appeared as a "discovery" in late 1951, became a "regular" in the next year, and by 1953 was receiving more fan mail than his boss. Godfrey, returning to the show in a bad mood after a hip operation, quietly fired a number of "lazy" staff members but gave LaRosa an abrupt on-air dismissal, saying to questioning reporters that the singer had gotten "too big for his britches." Although Alex McNeil's *Total Television* characterizes the firing as "one of the most widely publicized incidents in television history,"[2] younger viewers today would know nothing of it. However, nearly half a century later it was a useful memory fragment to one New York couple, as Enid Nemy relates the story in the "Metropolitan Diary" column of *The New York Times:* "Sidney and Esther Sanders were on the Long Island Rail Road when the conductor questioned their half-price senior tickets. They assured him that they were, indeed, seniors. 'I'm going to test you,' he said. 'Who fired Julius LaRosa?' Mrs. Sanders didn't hesitate. 'Arthur Godfrey,' she said. 'Only seniors would know that,' he said, and went on his way."[3]

The mid-1950s brought a sharp turn in popular musical taste and redirected a large portion of the young television audience to AM radio, where teens could "Rock around the Clock," as the title of Bill Haley's song urged them to do. Richard Smith saw the doom of *Your Hit Parade* in the emergence of rock and roll: "Trying to substitute 'How Much Is That Doggie in The Window?' for the latest rock tune proved hilarious and ultimately impossible, closing them down." Elvis Presley and the Beatles made much anticipated appearances on Ed Sullivan's Sunday evening television show, but Presley's network debut prompted a new edict: in his subsequent appearances, the studio audience could see what it might, but home viewers would see the young man from Memphis only from the navel up—no gyrating and thrusting hips. Sara Robertson illustrates the generational problem: "I recall visiting my cousin on a Sunday evening when Elvis Presley was to be featured on Ed Sullivan's show. We were all excited about seeing Elvis, but Daddy made us leave. He thought Elvis was the devil personified (I think he was totally clueless as to the rock and roll phenomenon)."

American Bandstand, evolving from a local Philadelphia show and coming to the ABC network with Dick Clark as host in 1957, was the music standard for many teenagers of the day. Guests danced to rock and roll tunes in a set decorated to suggest a record shop, and best-liked dancers paid many return visits. This was the "favorite program" of Ann Lewis, who was in junior high school then: "The program was on each afternoon at 4:00 P.M. We would rush home from school so we would not miss a single minute. We knew all the regulars' names, and we would talk about the show every day at school as if they were our best friends. We knew who was dating whom, and when they would break up, we would take it personally."

With Elvis Presley startling parental viewers on the Sullivan variety show and *Bandstand* institutionalizing the new music and dancing styles, a fracturing of popular music taste had clearly begun. The general variety shows would be victims of that change, and teen-specific offerings such as ABC's *Shindig* and NBC's *Hullabaloo* emerged in the mid-1960s. Beth Jarrard says that she and her cousin Mary Alice "and other high school friends became fans of the new teen music shows. We had watched *American Bandstand* for years, but *Hullabaloo* and *Shindig* had a different look and more energy—and they were on at night. When the national touring company of *Shindig* came to the Greenville [South Carolina] Memorial Auditorium, we bought tickets, and Mary Alice's older brother drove us to town. We felt really 'with it.'" Tom Maertens, Jr., confesses an uncomplicated attraction to the teen shows: "As I got older, I always liked to watch *Shindig* and *Hullabaloo* just to ogle the go-go dancers in the short skirts."

Television series such as *Hootenanny*, hosted by a different college campus each week, and *Hee-Haw*, a variety show for the older Nashville set, provided bales of hay for their performers and sometimes their camera-range studio audiences to sit on. Bob Fattibene was a student at the University of Maryland in 1963 when *Hootenanny* filmed two shows there, and as a member of the audience he "got considerable 'airtime.'" He searches the Internet today, looking for tapes of those shows hosted by Jack Linkletter.

Situation Comedies

"*Father Knows Best*," repeats Marty Duckenfield of the National Dropout Prevention Center, as she sees her own family experience reflected in the title of Robert Young's pioneering sitcom, which came to TV from radio in 1954. "My mother was an immigrant from Scandinavia, and she was eager to be a good American citizen, to do everything right. If *Father Knows Best* is what the American way says, then that was the way for her." Reeling off the names of the show's characters and of her siblings, she further notes that the program closely mirrored age patterns and other relationships in her immediate family.

Another family comedy of that era caused teenaged Beth Jarrard to face the issue of "a room of one's own." She recalls, "I grew up in a tiny town in South Carolina, and although my father and mother stressed the importance of education and travel—by the time I was seven I had visited 15 states—TV opened my eyes to many other places and things. I remember watching *Make Room for Daddy*, with Danny Thomas and Marjorie Lord, and wishing I could live in an apartment. I lived in a house with my own room and a huge yard with a swimming pool, but that apartment looked so exciting and different from what I had, and different often seems better."

"How many remember *I Married Joan*, starring Jim Baccus?" Suzanne Traenkle asks. "One of the funniest episodes I ever saw was when the husband and Joan [Joan Stevens, played by Joan Davis] were mad at each other. He sat at one end of the table and she at the other. To pass the breadbasket, they used tennis rackets to literally hit the bread back and forth. By the time the scene was over, they were laughing."

Norma Rodriguez recalls the 1950s as "a difficult time for my family since my mother was terminally ill and we younger ones bounced around from home to the aunt and uncle's house." She found consolation in the family sitcom *The Stuart Erwin Show* (later syndicated as *Life with the Erwins* and *The Trouble with Father*), set in a two-story house that affirmed the habitat "norm" of *Father Knows Best* and *Ozzie and Harriet*. "I remember Erwin as sort of a Tom Bosley-type person, the warm, wise father," she says. "Jackie, the younger daughter, I remember as a tomboy, sort of a Zelda from *Dobie Gillis*. Willy, the handyman, I remember as an African-American character actor of that era."

The title character in *The Many Loves of Dobie Gillis*, played by Dwayne Hickman under a preternaturally blond head of hair, constantly entangled himself in his schemes to attract girls, and his difficulties were compounded by the advice of his lazy friend Maynard G. Krebs (Bob Denver). "We learned much of our [teen] vocabulary from such shows," Beth Jarrard says, explaining, "Words such as 'hip' and 'cool' were popularized through TV, and I first became aware of the term 'beatnik' from watching *Dobie Gillis*. Maynard G. Krebs probably didn't bear any resemblance to the serious beatniks of the period, but I copied his outfit for a Halloween costume one year."

For teen girls watching *Ozzie and Harriet* during the later seasons of its fourteen-year run (1952–1966) on ABC, the other members of the Nelson family mattered little: perpetually puttering father Ozzie, kitchen-bound Harriet, and quieter brother David seemed mere backdrops to sharp-featured Ricky, whose early singing efforts were portrayed within the cozy sitcom. Ozzie Nelson had begun his show business career as the leader of a 1930s dance band. Guitar-wielding son Ricky was the stuff of wall posters and sweet dreams to a clearly different generation. Today Richard Smith sees this show as "important for reflecting the 'typical' '50s family as well as for developing Ricky Nelson. As he evolved as a singer (to rival Elvis for a time), he used to sing his latest hit at the end of the show to the squeals of adoring teenager girls."

Also appealing to female eyes was Robert Cummings, the somewhat older girl-chasing bachelor in *Love That Bob* and *The Bob Cummings Show*. Marilyn Chadwick declares, "Bob was the most handsome man I ever saw, and I resolved to marry a man some day that looked just like him. We loved his secretary, Shultzy, played by Ann B. Davis of future *Brady Bunch* fame." However, the original *Dick Van Dyke Show* brought a husband-model quandary for Marilyn Chadwick: "The entire cast rated an A-plus in every episode, but Bob Cummings now needed to blend with Dick Van Dyke as ideal future husband material! Mary Tyler Moore (as his wife) was also one of the first fashion trendsetters. She popularized the Capri pants and the capezio shoes with very pointed toes. Of course, no one could wear either as well as Mary, but scores of viewers did their best, myself included."

Other sitcoms earned the loyal viewing of Marilyn Chadwick and her family. Each show offered a hook of personal identification:

Private Secretary, starring Ann Sothern, kept us in stitches. Her job was a lot more fun than the secretarial position I held in later years. When I was ten years old, I would sit at our little writing desk that was built into the wall between our dining and living rooms and mimic all the actions and dialogue of Ann. It was about that time that I wanted blond hair just like hers, but my ever-so-sensible mother counseled me into being satisfied as a brunette.

I wonder how many people remember a program entitled *December Bride*. This was another winner, starring Spring Byington [as a widow] and Frances Rafferty as her married daughter. While this was in its height of popularity, my mother got wind of the fact that Frances had relatives in Sioux City and was coming to spend Christmas with them. Once again the four of us (plus dog) jumped in the car and embarked on an excursion past the home where the celebrity was staying. I remember feeling a little uneasy. It was nighttime, and I felt a little like a window peeker, but I quickly put together in my mind a little story about how we were out viewing the holiday lighting exhibits on these beautiful country club homes, just in case a policeman would find our behavior a little suspicious. All was worth the trip, however, when we actually saw a woman, who we insist to this day was Frances, slowly walk past the gigantic picture window. It was a thrill and a relief, because I was less apprehensive about being arrested now that our mission was more than accomplished.

Everyone seems to know who Lucy, Desi, Fred, and Ethel were. Lucy and Ethel reminded me so much of my mother and her best friend, Marian, who were "birds of a feather" and truly did everything together and had some pretty humorous tales to relate, just like Lucy and Ethel. Fred was one of a kind and didn't fit the usual supporting role. He looked exactly like every one of our next-door neighbors. His life and career ended way too soon. Many more years of him would have been a valuable gift to us all.

Corinne Holt Sawyer recalls her sister Madeline Campillo's viewing habits in the days of Lucille Ball's television prime: "She didn't go to night classes, go out to the movies, visit friends, allow friends to visit, or accept a date on Monday nights. Not ever. Because that was the night *I Love Lucy* was on, and she, like many Americans at that time, simply HAD to see *I Love Lucy* every week." Young Robert Moffat "couldn't miss *I Love Lucy*, even if the TV were broken—I would get Mother to arrange with a friend or neighbor to let me come over to see *Lucy*. [Since then] I've collected all of the *I Love Lucy* programs on tape."

Like *I Love Lucy* and the later *The Lucy Show*, CBS's *The Andy Griffith Show* earned one of the most devoted followings in TV history. Many Sunday school lessons, sermons, and commencement addresses and several devotional books have been based on the doings and sayings of the inhabitants of Mayberry, North Carolina, the show's representation of Andy Griffith's native Mt. Airy. Ned Willey, commenting on his daughter's love for the show, asks rhetorically, "Why do you think she named her son 'Andy'?" Ronnie Howard as freckled Opie, Don Knotts as the bumbling deputy to Sheriff Andy, Jim Nabors as the service station man, and Frances Bavier as Andy's kindly Aunt Bee proved to be irresistible characters who have helped *The Andy Griffith Show* outlive the other rural comedies of the 1960s and early 1970s (*The Beverly Hillbillies, Green Acres, Petticoat Junction*) in syndication. Vickie Metz notes, "My husband and I are both avid *Andy Griffith Show* fans. We have seen the episodes so many times, and we share many private jokes taken from the dialogue of the shows. He even proposed while we were watching an episode of *The Andy Griffith Show!*"

Columnist Rheta Grimsley Johnson, watching series reruns during her recuperation from a case of flu in 2000, ruminated over *The Andy Griffith Show:* "I once considered moving to North Carolina because in that state you can turn on the TV any time of the day or night and on some channel find Andy Griffith. Like any true Andy aficionado, I can quote verbatim from most episodes, which are so loaded with folk wisdom and decency you wonder how the show ever stayed on the air past the first week. (Compared to a trash show like *Married with Children*, Andy Griffith's humor seems pure and profound.")[4]

In another of CBS's "rural comedies," *Green Acres*, the city slickers move to the country and find, in place of the anticipated peace and quiet, loudly and farcically complicated situations in Hooterville. In 2003, Sherri Maxman, contributing a "for the 35-and-over crowd and fans of nostalgia TV" item to the weekly column "Metropolitan Diary" in *The New York Times*, showed how a seed planted by that 1965–1971 program finally popped into fruition: "On a recent Saturday evening, we attended a Brooklyn Cyclones game with a group of friends and then took the subway back to Manhattan. As we emerged from the steamy underground at 42nd Street and Seventh Avenue, our friend Liz proclaimed, 'Fresh air—Times Square!' She grinned and added, 'I've always wanted to be able to say that.' Of course, we had to explain to our 9-year-old about the *Green Acres* theme song after we all fell about laughing."[5]

An early Halloween-night episode of CBS's Appalachia-gone-West comedy *The Beverly Hillbillies* produced a crisis and prompted a solution for Kathy Chastine-Burrell's children: "We took the three oldest kids out trick-or-treating for a short while and came back home to put the kids to bed so we could watch the show. The kids were so disappointed [that the program was scheduled beyond their bedtime] because we had looked forward to this for weeks. I found out the next morning that Steve, Elaine, and Mike had somehow made an opening in their bedroom door and watched the program anyway!" She adds, "I could not tell you anything about the new shows today, for I never see them. I am too busy surfing the channels for *All in the Family, The Golden Girls*, and *I Love Lucy*."

Indeed, for viewers who grew up with *Father Knows Best*, the *Lucy* shows, and *Andy Griffith*, later network sitcoms lacked a sense of warmth. Julie Dunlap found mere "silliness" in the comedy shows of the 1980s and "disgust" for commercial network shows "through the '90s and to the present." Dixie Ann Schmittou adds, "I'm really sorry that network TV programming has deteriorated so. I loved all Bob Newhart's shows except the last one. They were genuinely funny without using crude language and sexual innuendo. I still laugh about Larry, 'his brother Darryl, and his other brother Darryl' [on *Newhart*]. I used to kid my sister Betty, who also had a daughter named Betty, and would jokingly introduce them as 'Betty and my other sister Betty.' Just the other day I had a phone call from my niece (they live in Michigan), and she said, 'This is the other Betty.'"

Beyond Comedy: Other Drama Programs

Popular memory holds that the first full decade of network telecasting was a "golden age" of TV drama. National Educational Television, predecessor of PBS, contributed strong productions, and on the commercial networks, weekly and biweekly anthology series offered live sixty- and ninety-minute original plays and adaptations of literary classics. Establishing corporate prestige often seemed to override the selling of specific products in the sponsorships of productions adapted to the small screen by well-known Broadway and Hollywood directors and performers. *Kraft Television Theatre* had had a brief run for New York viewers in

1940, and in 1947 the program came to the NBC network (with a brief overlapping run on ABC in 1953). In the next decade it presented more than 600 productions starring Jack Lemmon, Anne Bancroft, Maggie Smith, Earl Holliman, Lee Remick, James Whitmore, and similarly well-known actors and actresses. Alcoa Aluminum offered filmed dramas under the series titles *The Alcoa Hour, Alcoa Theatre*, and *Alcoa Premiere*, the latter hosted by dancer Fred Astaire. Alternating Sunday evening presentations with *Philco Television Playhouse* and later with *The Alcoa Hour*, the *Goodyear Playhouse* offered such productions as Paddy Chayefsky's *Marty* starring Rod Steiger and Nancy Marchand, Gore Vidal's *Visit to a Small Planet* with Cyril Ritchard, and Chayefsky's *The Catered Affair*, starring Thelma Ritter. Kathy Chastine-Burrell remembers "especially" an early live production on *The Hallmark Hall of Fame:* "The show was *Green Pastures*, and it was an all-black cast. I have never seen one since [which was] that good." *Playhouse 90* belongs on even the briefest list of prestigious drama anthology programs, and retired New York policeman Richard Smith suggests a reason: "One of the attractions of watching this program was knowing that it was live—hence the possibility of some dramatic screw-ups. Even Tab Hunter, that heartthrob of the era, acted on *Playhouse 90*, performing as a cat burglar to creditable reviews."

From 1953 until 1961, film star Loretta Young presided over a Sunday 10:00 P.M. half hour anthology of human interest stories, starring in many episodes herself. Richard Smith was among the many impressed by the attractive host whose entrance, very different from Milton Berle's, was equally memorable: "With her svelte figure (reminiscent of today's Calista Lockhart type) she had a beautiful face with high cheekbones. Interestingly, she modeled a different outfit each week as she 'swirled' through the door to begin the program. Loretta Young epitomized the feminine elegance of the early television era."

Lavishly produced general drama anthologies began to disappear from television in the late 1950s, while two especially well-hosted new shows specialized in frightening plots, often with surprise endings. As host of *Alfred Hitchcock Presents*, the British-born film director assumed the ghoulish whimsy of a figure in a Charles Addams cartoon. Hitchcock's full lower face, supported by a generous double chin, shaped itself toward a deceptively childish oval mouth as he delivered his opening teasers and closing comments, which often depended on a deadpan enunciation of outlandish puns. "The next day at school or work people remembered Hitchcock's macabre jokes better than the plots, oftentimes," Alton Landers says. For his science fiction series *The Twilight Zone* host Rod Serling typically wrote or borrowed stories that placed their characters on the "other side of things," disturbingly thrust into time-warped shifts of place and identity. Alton Landers remembers, "Every week that show scared all the knots out of my pajamas." Four decades after the end of its original run in 1964, Serling's program was a strong memory for Susan Lees, who tells of being called to jury duty in a civil case brought by a woman who had fallen on an ill-maintained sidewalk and broken her wrist. During the jury's lunch break Susan Lees herself fell and broke her wrist "at that same spot," and while waiting for the ambulance, she witnessed a man's falling at the same uneven place on the sidewalk. Her conclusion: "Kind of *Twilight Zone*, no?"[6]

West Coast movie studios had initially held out against television, thinking that the new medium was merely a fad. H. M. Warner had said of sound films in 1927, "Who the hell wants to hear actors talk?" and Twentieth Century-Fox mogul Darryl F. Zanuck was equally shortsighted when he asserted in 1946, "Television won't be able to hold on to any market it captures after the first six months." Movie and TV interests fought bitterly until they realized that mutual benefit could be more profitable than sustained rivalry. As "Golden Age" anthology dramas began to seem more and more suffocated by the lighting and space constraints of the TV studio, Hollywood production facilities came available for prime time Western shows and police or detective series. Keeping the peace through decisive action supplemented by clipped dialogue, Bat Masterson, Wyatt Earp, Marshal Matt Dillon of *Gunsmoke*, and Paladin of *Have Gun, Will Travel* did for the nineteenth-century Western Territories what Jack Webb (as Sgt. Joe Friday) did for twentieth-century Los Angeles in *Dragnet*. Robert Stack starred as Treasury agent Eliot Ness in the Prohibition-decade crime show *The Untouchables*, while *The Roaring Twenties* featured Dorothy Provine as a most attractive flapper. "I had a huge crush on Dorothy Provine (even bigger than the one for Gail Storm on *My Little Margie*)," says Art Donahoe, adding, "The show wasn't on long [two seasons on ABC], but it provided material for adolescent fantasies."

The detective show *Peter Gunn*, starring Craig Stevens, centered much of its action in a seamy side drinking establishment ironically named Mother's. With its jazzy score by Henry Mancini, this was one of the shows that gave Beth Jarrard

> other lifestyle impressions, many that stayed with me for years to come. The characters of *Peter Gunn* were so sophisticated, and all the women wore beautiful clothes. These sophisticated people and others in popular dramas often had drinks in their hands. My family didn't drink. In fact, the only people we knew who did were the more downtrodden citizens in our town. As a preteen, I recall removing from the china cabinet in the dining room one of my mother's crystal "sherbert" glasses that looked suspiciously like the barware I often saw on TV, and I filled it with Coca-Cola for a bit of play acting. I did not become an alcoholic or even a regular drinker in later years, but I can vouch for the influence these kinds of images have on children.

ABC's *The Fugitive*, starring David Janssen as Dr. Richard Kimble, a man wrongly convicted of murdering his wife, offered a fresh variation on the vigilantism of some Western series. As the central character spent four seasons attempting to clear his name, this became another favorite series in Beth Jarrard's home: "We watched the entire run of *The Fugitive*, wondering if they would ever reveal the identity of the one-armed man who killed Helen Kimble. We cheered for Dr. Kimble in his run from Detective Gerard, wishing this sinister character could have had another name. Ours is spelled 'J-A-R-R-A-R-D,' but it is pronounced the same."

"Saturday nights were a pretty lively and busy time in the Jarrard household," she continues. "Even though Mom and Dad worked six days a week at the store and were tired, Saturday nights meant early preparations for Sunday dinner and clothes to get ready for church. But everything stopped when *Perry Mason* came on. It was one of my mother's favorite shows, and to this day she watches any time cable stations rerun the old episodes or the later movies in which Raymond Burr reprised that role." Raymond Burr dominated the courtroom scene as Erle Stanley Gardner's criminal lawyer from 1957 to 1966 and returned to similar premises as the wheelchair-bound but intellectually adept survivor of an assassination attempt in *Ironside* from 1967 to 1975.

Although it echoed Cold War international intrigue themes and Army-McCarthy hearings issues of the 1950s, NBC's *The Man from U.N.C.L.E.*, starring Robert Vaughan and David McCallum, found an agreeable lightness in its 1964–1968 run. Sherri Butler discerned useful "lessons" in it:

Intended to capitalize on the success of the James Bond movies, it depicted two spies saving the world in exotic locales week after week. It was worth viewing for this round-the-world aspect alone, even if all its exotic locales were really MGM backlots. But there was more. The creator, Sam Rolfe, chose to be a little bold and risky for the mid-1960s. He envisioned a global organization in which all nations cooperated and made one of his heroes a Soviet, even showing him, briefly, in a Soviet military uniform. So, in the middle of the Cold War, with the U.S. and Soviet Union bristling at each other, American kids were rooting once a week for an American-Soviet team, Napoleon Solo and Illya Kuryakin, working together to conquer evil.[7]

In the same years NBC's *I Spy* made a significant dent in the color line by teaming Bill Cosby (later to earn even greater fame as the star of a family sitcom) and Robert Culp as widely traveling undercover agents.

Sherri Butler watched another dramatized "travel" series with a mixture of curiosity and disbelief: "*Route 66* took viewers across the country with two singularly sappy guys," played by Martin Milner and George Maharis, the latter replaced by Glenn Corbett as the codriver of the touring Corvette in the 1960–1964 NBC series. "But," she continues, "the background to all the stories was the great American highway. Wonder how many folks were introduced to this historic American road and the land it crossed through this silly show?"[8] Personal experience led Joanne Wingard to a very different view of that program and one of its stars: "In 1960, *Route 66* was filming in Pittsburgh," she recalled in 2001, and she took her home movie camera to the courtyard of a downtown office complex, hoping to get some footage of George Maharis in the scene being recorded thirty to forty feet away from her vantage point. Although the crew asked her to stop filming because the clicking of her camera might be picked up on the production sound equipment, she "met George Maharis, and we became fast friends. He is the godfather to one of my children, and I could write a book of anecdotes about him and about going all over the country to see him in many, many plays after the series was cancelled." Her son Dan adds that Maharis "came to our house a few times, and my mom and dad were his guests in New York City a couple of times. I met him four or five times as a kid, but my mom kept in touch with him until she died" in January 2004.

Against the growing social and political unrest of the 1960s, patriotic and heroic themes were played out in *The Gallant Men, Twelve o' clock High, The Rat Patrol,* and other World War II dramatic shows. The most durable of these was *Combat,* running on ABC from 1962 to 1967 and well-remembered by the Rev. Andy Oliver:

> Playing army was done by the honor system in our neighborhood. This was before flashing lasers and multi-colored paintballs. You were dead or wounded when simply declared that way, and you were healed or resurrected at your own discretion. Our battlefield was set amidst split-levels and brick ranchers, and our tactics and roles were determined by our favorite WWII-based TV show, *Combat!*
>
> The real battle in our neighborhood was actually who would be which *Combat* character. There were few choices. You could be the hard-nosed, cold sergeant; the compassionate, somewhat caring lieutenant; or one of the less prominent, almost nameless other soldiers who were subject to mortal wounds and even death on any week's episode of our favorite war drama. It was each suburban soldier's responsibility to die or be wounded as dramatically as possible, and often our words were littered with phrases from that particular week's episode. The only phrase I remember, however, is one that the heartless sergeant, [played by] Vic Morrow, used on a bewildered and frightened infantryman. The terrified soldier shared his doubts and fears with Morrow about a certain objective. Let's just say he was ordered to destroy a bunker. He tells the sergeant, "I'll try, sir." I can see Vic Morrow now with steely eyes and clinched jaw replying, "Don't try. You do it!" He was cold, but he sure knew how to get the job done. There's no telling how often I told frightened neighborhood warriors when I assumed the sergeant's role, "Don't try. You do it!"
>
> "Sergeant, I'll try to take the tree house."
>
> *"Don't try. You do it!"*
>
> "Sergeant, that front porch is well fortified. I'll try."
>
> *"Don't try. You do it!"*
>
> "Sergeant, I think it's impossible to take the culvert. I'll try."
>
> *"Don't try. You do it!"*
>
> Playing army is a thing of the past now. I guess it's rather "incorrect" for today's kids.
>
> However, *Combat* will always remain a part of my life. The other day my oldest son told me he would "try" to improve his ailing biology grade. At once I felt somewhat conflicted. Should I be compassionate and understanding like the sensitive lieutenant or take the hard line like Old Sarge? I took the merciful approach toward my shell-shocked youngster. If that tactic doesn't work, however, I'm afraid I'll have to employ another strategy. I'll have to hop in the trench with him, look him in the eye, and coldly tell him, *"Don't try to improve that grade, son. You do it!"*
>
> I'm still fighting battles; only the neighborhood and enemies have changed. I'm thankful for my *Combat* training.

Hollywood B-Westerns were prominent in the "kiddy schlock" that filled after-school hours and Saturday mornings on early TV. Veteran plains-rider William Boyd proved a good businessman in buying the television rights to his many theatrical Westerns and reediting them for half- and full-hour small-screen presentation. Newly produced thirty-minute shows featured movie cowboy rivals Gene Autry, assisted on his Melody Ranch by giggling Smiley Burnett and others, and Roy Rogers (sometimes riding Trigger and sometimes riding in Pat Brady's jeep Nellybelle) and wife Dale Evans (riding Buttermilk) with assorted sidekicks, including the Sons of the Pioneers singing group. The Lone Ranger and his Indian companion Tonto, radio favorites since 1933, rode Silver and Scout respectively into the television picture in 1949. The durable Western anthology *Death Valley Days*, begun on radio in 1930, ran as a syndicated TV series from 1952 to 1970.

However, a miniseries starring Fess Parker as frontiersman Davy Crockett, who died at the Alamo, was one of the Frontierland features in the first season (1954–1955) of ABC's Wednesday evening *Disneyland*, and that fresh treatment of nineteenth century law-and-order themes prompted not only a several-seasons fad of prime time Western programs but also landmark sales of coonskin hats in imitation of Crockett's headgear. "That was the one night a week when everyone was allowed to stay up an extra hour to watch," Dennis Thompson says, adding, "There were so many coonskin caps and Davy Crockett lunchboxes in the elementary school when the Davy Crockett series aired!" Soon Western dramas dotted the TV schedules like cacti around Tucson. *Gunsmoke*, having started on CBS radio in 1952 with William Conrad as Marshal Matt Dillon, was readied for its television debut in September 1955 with "that tall drink o' water" James Arness in the lead role. The offscreen doings of Long Branch saloon owner Miss Kitty (Amanda Blake) and lame deputy Chester B. Goode (Dennis Weaver) hinted at adult themes thought to be over the heads of those younger viewers who were still awake at the show's Saturday 10:00 P.M. time slot. Tallness and a quick shooting arm were among the credentials that Chuck Connors brought to the title role in ABC's *The Rifleman* from 1958 to 1963. Reserved seriousness governed Richard Boone's 1957–1963 CBS portrayal of San Francisco-based Paladin, whose business card offered, "*Have Gun, Will Travel*." Others—*Bat Masterson* (1958–1961, NBC), *Broken Arrow* (1956–1958, ABC), *The Life and Legend of Wyatt Earp* (1955–1961, ABC), *Cheyenne* (1955–1963, ABC)—provided variations on the shoot-quickly, talk-later model of evening "adult" Westerns. A shift in emphasis from the confrontations of the frontier lawman to the difficulties of the cattle rancher came to the TV Western genre in 1959 through NBC's extravagantly produced *Bonanza*, starring Lorne Greene as Ben Cartwright, a rancher with three sons (each by a different wife), countless cattle, and plenty of Ponderosa acreage to oversee (until 1971). From 1965 to 1969 ABC's *The Big Valley* brought a feminist perspective to the frontier, with Barbara Stanwyck as a widow who sought to maintain her California ranch with her four sons and a half-Indian stepson. Inevitably the Western genre was parodied, most successfully by *F Troop*

(1965–1967, ABC), where the cavalrymen and the Indians were equally inept in pursuing their opposed interests.

For a decade and a half, then, viewers had many choices of Western melodramas, and program loyalties were often deep. Joan Henry says, "My fondest memories were in junior high school when I would sneak down to the basement (the TV was still not allowed in the living room) and watch *Rawhide*. Rowdy Yates (Clint Eastwood) was my first crush. I still have a crush on him!" Young Liz Newall, today the editor of a university alumni magazine, was so taken with *Gunsmoke* that she began writing her own scripts for the show. Amy Blackmarr, in an autobiography describing her growing up in Ocilla, Georgia, and her decision to return to that area years later, says, "It's funny that I grew up in love with Roy Rogers and Dale Evans and horses in particular, and Westerns on television like *Laredo* and *The Virginian* and *Bonanza*, because my family never owned any horses and [when offered one as a gift] my folks wouldn't have any part of that. . . . I had a crush on Trampas [played by Doug McClure in *The Virginian*] but Hoss Cartwright was my dream husband. When I was a girl my first wish was to marry a Texas rancher. He'd be just like Hoss and we'd own a string of horses and have land that stretched as far as I could see like on *High Chaparral*."[9]

Growing up in Tomah, Wisconsin, catching tadpoles and learning to shoot rifles, David Benjamin intuited that boys were meant to live by an unspoken Code hinted at in the theme music and the premise of *Have Gun, Will Travel*:

> For grownups, *Have Gun, Will Travel* was a TV show on Saturday night, in between *Wanted: Dead or Alive* and *Gunsmoke*. For kids in Tomah, that series of three Westerns on Channel 8 on Saturday was holier than the Stations of the Cross. *Have Gun, Will Travel* was our Apostles' Creed. Every kid knew the look and contents of that card, which Richard Boone, starring as Paladin, flashed in every show. We didn't know that "paladin" was a synonym for "knight," nor did most of us grasp the symbolic interplay of word and image. None of us perceived any irony in the fact that we had no guns and most of us needed permission to "travel" across Jackson Street. What we knew was that every one of us was a Paladin in our soul. We lived in the apparent comfort of home and hearth, as Paladin lived in the opulence of Frisco's fanciest hotel. But at a moment's notice, each one of us might be cast (metaphorically) into the wilderness, thrown from our horse, disarmed and bleeding from a bullet in our thigh, drinking only the water we could suck from a barrel cactus, shaking a fist at the circling vultures and hunted by villains so murderous that no ordinary lawman dared stand up to them. . . .
>
> Every kid knew every word to the theme song. Every kid I ever knew was "a knight without armor in a savage land."

As further proof that this childhood epiphany was truly a redefining moment, David Benjamin uses glancing references to Paladin's card and to the show's theme lyrics as a unifying element in his winsome autobiography *The Life and Times of the Last Kid Picked*.[10]

In Richard Woodward's 1950s neighborhood in West Knoxville, Tennessee, the houses formed a circle (much like camping configurations of the wagon trains in Western films and TV series), and one of those houses had a television set. Every afternoon at 5:00 P.M. all of the boys gathered there to watch *The Lone Ranger*, drawing their toy guns and using the back of the couch as a strategic shooting position. Elsewhere, Mary Eberhart watched alone but not passively: "When I was young, *The Lone Ranger* was my favorite show, and Tonto was my favorite character. One day my mother heard a crash and then crying. When she investigated, she found me on the floor behind the couch, scratched from the 'textured' wallpaper and bruised from the back of the couch. When she asked me what in the world I thought I was doing, I informed her, 'Tonto was chasing me!'" Barbara Bergman too came to that program with a very impressionable understanding: "I was born in 1953 and was just a young 'un when we got our first TV set. My favorite shows were *Sky King* and *Rin Tin Tin* and especially *The Lone Ranger*. I loved to sing his song in kindergarten, 'Away with the Ranger.' It wasn't until I was in first grade that I learned that the words were supposed to be 'Away in the Manger.'"

For Priscilla Kanet and her sister Caroline Jasper, TV Westerns were invitations to dress-up activities, and they found female role models there. Caroline, today an artist who sells many of her paintings on the Internet, counts off their favorites, beginning with *Tales of the Texas Rangers:*

Priscilla and I never missed it. As soon as the theme song started, the two of us jumped up and pretended to ride our horses around in a circle in front of the TV, loudly singing along.

"Davy Crockett" on the Walt Disney television show may have launched the whole TV memorabilia business. Davy Crockett coonskin caps with tails at the back were everywhere. I had a pastel yellow one, and I think Priscilla's was pink. There were normal looking faux fur caps for boys. The theme song record, on one of those colored plastic 45s, dominated use of our portable record player.

All of the Westerns were popular. We loved *Hopalong Cassidy* and especially *Annie Oakley* and *Roy Rogers*. Dale Evans, Priscilla's alter ego, and Annie Oakley, mine, were constants in our daily play activities. We did our best to recreate their costumes and made up our own scenarios to play out with neighborhood kids. We lived on a farm, and the [nearby] kids with the first TV also had ponies. We were serious cowgirls! [I still have] a snapshot of me as Annie Oakley, age 5, in pigtails, wearing a straw close-as-I-could-get-to-Western hat, jacket zipped up inside-out to show the plaid lining, pony print fringed scarf (I still have it), cowboy boots (actually just rubber boots printed with a Western design), and displaying my lasso. I loved getting dressed up like that.

Annie Oakley was my idol. Long after the dress-up days, another experience made for an unhappy memory when I got to meet the star [Gail Davis] who created the TV role. About 25 years later, I discovered her running a pony ride at a little country fair in Georgia. When I tried to tell her how much her show meant to me as a kid, she made me sorry to have approached her. She was bitter and unfriendly. Children riding her tired ponies didn't seem to notice.

Television didn't seem so far from reality when I was five.

Priscilla Kanet's recollections amplify her sister's:

We grew up in rural Maryland, farm kids, and TV became part of our play—Wild West games, bank robbers, etc.

Cowboy hats and outfits, Polly Crockett (not Davy—even then, women had a role) were regular parts of our attire. My Polly Crockett hat was really cool. We "had" to buy them at Stuckey's [gas-food-souvenirs tourist stop] on one of our annual summer trips to South Carolina. It was yellow fur with a pseudoleather top. I remember putting a doll to sleep with it, using the tail as a cover.

My dad loved *Gunsmoke*, and watching it was a family ritual. I guess he had good taste; 25 years is a good run [for a TV show]. I learned about Wyatt Earp and Bat Masterson from TV too.

The "great Saturday night Westerns" *Have Gun, Will Travel* and *Gunsmoke* were regular viewing fare in Beth Jarrard's home: "My younger brother Rick had a set of toy 'Paladin' pistols, and he was not the only kid in our town who took to mimicking the stiff-legged walk of Dennis Weaver in the role of Chester, and who ran around yelling, 'Mis-ter Dillon, Mis-ter Dillon.'"

Like the Westerns, soap operas bubbled up into evening schedules from their familiar daytime slots. A 1964–1969 serial adaptation of Grace Metalious's sensational novel *Peyton Place* started a trend that led to *Dallas* (1978–1991) and other bathetic and often pulchritudinous series.

An unexpected encounter with daytime serials led Skip Eisiminger to declare television a "forked medium," of which the soaps are "the left prong" and commercials the right. He recalls,

When I was a graduate student at Auburn University in the late '60s, I came home a bit earlier than usual one Friday afternoon to our duplex in Columbus, Georgia. As I walked into the darkened living room, there squatted our elderly babysitter on the carpet, tears on her cheeks, a wadded tissue in her hand, watching *As the Stomach Turns* or some such soap opera. The house my wife and I were renting had a door that allowed the owner, our sitter, access to both sides of the house, so it wasn't unusual to find her in our apartment. On this occasion when her set ("the black hole of Calcutta," she called it) had blown a tube, she let herself in and took a seat on the carpet, thinking I'd be home at 4:30, not 3:30. This widow hunkered on the carpet (so as not to "mess up" our couch) was a pathetic sight, but the venerable soaps helped to fill a void in her life that neither our small son, her grown children, her grandchildren, nor the church satisfied. I told her to take a seat on the couch and watch all she cared to.

Beth Jarrard knew the soaps in both their afternoon and their evening manifestations:

After I left home, *Dallas* and *Falcon Crest*, two popular nighttime soaps, became a topic of conversation during our long-distance phone calls. Mother, an avid movie fan since growing up with a young widowed mother in Greer, South Carolina, loved seeing Jane Wyman as the matriarch of Falcon Crest vineyard, and we both offered up theories on the solution to the *Dallas* cliffhanger question "Who shot J.R.?"

And speaking of soaps, I watched *Another World*, off and on, from its very first day in 1964 and throughout its 35-year run. My viewing began as a pastime while waiting for my ride home from school. My Aunt Millie was a teacher and lived on my street. Her daughter, Mary Alice, our friend Linda Guest, daughter of the science teacher, and I watched TV in the science lab while waiting for the end of the teachers' day. Later we added *Dark Shadows* and *Days of Our Lives* to the schedule. I'm sure Mrs. Guest would have preferred we use the time to study the periodic tables on the wall, but we preferred the soaps.

Marilyn Chadwick picked up decorating ideas for her room from sitcoms and from daytime serials. "Speaking of soaps," she says, "I rarely saw those early ones, unless I was home sick from school. *Search for Tomorrow, Love of Life*, and *Guiding Light* were the ones I remember. Their characters were vivid to me, though. JoAnn Tate (*Search*) and Vanessa Sterling (*Love*) spent a lot of time in the kitchen wearing their aprons, baking cookies, and solving problems. This homemaker stuff appealed to me, especially since the children were always 'playing in their rooms' or 'taking a nap.' I picked up an inaccurate impression of motherhood as being relatively simple."

In the days before other programming was readily available on many channels, movies filled large swatches of airtime. In the mid-1950s in New York City, Allan King remembers, "we had a show called *The Million Dollar Movie*, which aired the same movie twice nightly, Monday through Friday at 7:30 P.M. and at 11:00 P.M. Saturdays and Sundays, the movie was repeated back to back to back to back. This was before the advent of sports coverage as we know it now." He adds, "My favorite was *Chiller Theater* on Saturday nights. Nothing like scaring the kids a little before sending them off for a good night's rest."

WOR-TV's *Million Dollar Movie* was customary fare for Marty Duckenfield's family too, although strictly observed weeknight bedtimes meant that "we stopped watching at the same point in the film every night," and they did not see the end of the week's featured movie until the weekend repeats. They were especially drawn to Fred Astaire and Ginger Rogers' 1939 RKO biopic *The Story of Vernon and Irene Castle*, and each evening they enjoyed the cheerful opening reels about the ballroom dancers' early careers. Finally able to see the remainder on one of the Saturday showings, they were "devastated" to learn that Vernon Castle had died in a plane crash.

Children's Programs

Clearly children eavesdropped on evening programs meant for adults, and concerns in that direction led to Family Hour restrictions, later eased or fudged, on prime time program content. Today, local stations' promos invite parents to note the availability of programs finely tuned to the educational, social, and emotional understandings and sensibilities of their viewing children. In earlier TV days, some stations carefully cultivated the young "pardners" who viewed the genially hosted B-Western movies, but other channels merely saw afterschool weekday and Saturday morning as programming segments to be filled as cheaply as possible with old feature films (chopped to fit time segments), cartoons, and serial chapters. Hollywood comedy shorts became the stuff of schoolground play, and Dennis Thompson remembers "my elementary school sending home notices asking that students not be allowed to watch *The Three Stooges* because so many of us were imitating their antics at school." (Those antics frequently involved farcical finger-in-the-eye and head-bonging routines.) The development from the short-lived DuMont network's 1949–1955 space saga *Captain Video* to CBS's *Captain Kangaroo*, with Bob Keeshan playing the large-pocketed avuncular host from 1955 to 1984, represents a revolution in thinking about programming for little angels, moppets, and rug rats.

Travis Montgomery Tucker, "Monty" to his friends, has been the president of a local Storytellers Guild and has often unreeled his tales in public gatherings. According to a human-interest column by Jeanne Brooks, he

> grew up hearing the adventures of *Sky King* and *The Lone Ranger* and *The Shadow* on radio. Kids now can have a hard time grasping that. Their question: "How did you know what was going on?" Tucker answers, "Oh, I could see everything."
>
> And eventually, when he did watch his favorite programs on a TV screen, "I was disappointed." He thought, "That's not the way that guy looks. That's not the way his ranch looks." Tucker says, "Today, they show you too much." And when they show you too much, they steal the story away from you and your imagination. Here you have the flaw of television and Hollywood. Even the best movies and TV shows contain the imagination in a box that is the director or producer's idea of what is beautiful or scary or tragic or thrilling.

That is why Travis Montgomery Tucker has spent many years telling stories to audiences assembled in bookstores and to a few eager listeners around a campfire.[11]

Carl Sturiale, also disappointed in the production values of early TV, remembers that the local *Space Funnies* show was so clumsily directed that in one episode "a guy's hand [was seen] adjusting the rocket ship which was in outer space. Talk about a long reach!"

The Stone Age cartoon series *The Flintstones* provided one of Germaine T. Leverette's strongest childhood memories:

> This show came on once a week in the evenings just like the sitcoms do now. It was my dad's favorite show. I still remember how everyone raced home from baseball games and other things in order to be home in time for Fred and Wilma's baby to be born. The pregnancy was a drawn-out event just like a soap opera, and it took a few

weeks for Pebbles to be born. We had many discussions in school, at the grocery store, at church and just about everywhere else we went, debating about whether the baby would be a boy or a girl. I don't remember getting much schoolwork done following the momentous birth of Pebbles.

The Mickey Mouse Club, running on weekday afternoons from 1955 to 1959 on ABC, wove together guest stars, newly produced serials, and the general cheer of the scrubbed and enthusiastic Mouseketeers. Dennis Thompson says, "The whole family dinner hour and schedule on weeknights had to be built around *The Mickey Mouse Club*, especially when the *Spin and Marty* serial was shown. I think every preteen boy in America at the time had their first crush on Annette Funicello!" As viewers grew along with the show, their perceptions of it changed, as Robert Cox discovered: "Although I was a teenager, I felt a bit like a dirty old man leering at all those little girls dancing around in their little black skirts on *The Mickey Mouse Club*. Like most young boys of the time, I had a thing for Annette [Funicello]. She was a decent, wholesome girl. It's a shame what happened to her" when she was later diagnosed with multiple sclerosis.

"Clarabel the clown came to our elementary school once," Marty Duckenfield says, smiling at the pleasant memory of the *Howdy Doody* character's contribution to her New York childhood. Dennis Keyes had an even better deal: he lived with the elf Mr. Jingeling and a number of other Cleveland TV characters, because his father played all of them. The son's recollections provide insight into the production conditions and even the economics of such shows:

I very vividly remember the days of LIVE television, especially when the local stations had to do much of their own programming. My father, the late Earl W. Keyes, worked for WEWS Channel 5 TV in Cleveland for many years. This was in the days of black and white and the old bulky roll-around TV cameras. He was the programming director for several shows, which originated locally from that station. One was a morning exercise program called *The Paige Palmer Show*, which aired about 9 A.M. One segment they had planned was a first aid demonstration, for which they needed a subject for mouth-to-mouth resuscitation. Dad asked if I would be able to go to the studio and be the "victim." I said, "Why not?" I had been interested in his work for a long time, so I went.

Another show he was involved in was the noontime kiddies show called *The Captain Penny Noon Show*. It featured a railroad engineer-type character who sort of emceed the showing of the Little Rascals cartoons. Between segments of the cartoon they either showed locally produced commercials from another live studio or played commercials from the projection room from vendor-supplied films. My father played the train ticket taker, Wilbert Wiffenpoof, wearing a costume, which he designed himself with dangling stuff from his cap and ticket stubs from everywhere stuck in his cap. This was in fleeting moments when he stepped from the control room, since he was also the director of the show. He also created a character for that show called

Mister Nicklesworth who was never seen, only heard as a voice, which he interjected from the control room booth.

Another more notable character that he portrayed in the early days of TV was Mr. Jingeling, one of Santa's elves, who was the keeper of the keys to all the toys that were wind-up type. The character was originated by another company and portrayed by Max Ellis until he died suddenly. Then my father adopted the role and portrayed it for many years, right up to the year before he died. He even secured the total rights to the character and the name after all the participants from the original advertising agency either passed on or released the rights. At first the stories he told on the show, which aired about 5 or 6 P.M., were specifically written about particular toys being sold at the Halle Bros. Store. Later the stories were about different Christmas themes, problems at the "toy factory" at Santa's North Pole, and finally about characters which were developed through the collaboration of my father and mother. Early on, the shows were either filmed or done live. Later they were prerecorded on 2-inch wide videotape and copied and distributed to other TV stations for later broadcast.

One day when I did not have school, Dad even let me spend the day with him at the studio. I sat with him in the director's control room and was even given the chance to "call the shots" for a small segment of one of the live shows. I even helped one Saturday afternoon in the splicing room editing the commercials into the afternoon movie reels.

Commercials

Those miniature dramas, television commercials, promised to make life simple, beautiful, exciting, more colorful, and less painful, and many a product's packaging and print advertising proclaimed, "As seen on TV!" Almost any product or service might be designated a "special TV offer" for added cachet. Swanson introduced TV dinners, and families bought collapsible leg-catching TV trays from which to eat them while facing TV-set-wards. TEE-VEE Records sold compilation LPs through the tube, and many other companies proclaimed that their products were "not sold in stores," thus underscoring the selling power of television. The infomercial was born to promote "amazing" multiuse "why-didn't-someone-think-of-that-before?" products for the kitchen (juicers, dicers, slicers, ricers) and the home workshop (instruments far more multiform than the renowned Swiss Army Knife). In *Tonight* show sketches Johnny Carson and his Mighty Carson Art Players frequently parodied the high volume Los Angeles car dealer, Cal Worthington, hawking his wares in flashy commercials and the seedy afternoon TV movie host constantly interrupting the film with a seemingly endless spiel of "brief messages." Dancing cigarette packages suggested the "lift" that a smoker might get from puffing at the promoted brand, and athletes (playing their own public images) and doctors (played by actors) argued the calming health benefits of "so mild" filtered brands. An aspirin commercial's irritating animated visuals of lightning-striking pain, accompanied by cymbal crashes, were enough to give any viewer a headache. Certs commercials featured vociferous arguments—"It's a CANDY mint!" "No, it's a BREATH mint!"—resolved in the all-knowing announcer's loudly repetitious closing proclamation that the product is in fact "TWO . . . [sound effect: CRASH] . . . TWO . . . [sound effect: CRASH] . . . *TWO* mints [sound effect: BOOM] in ONE!"

"I myself know prodigious numbers of TV program theme songs and commercial jingles from the '50s and '60s," says Computer Software Engineer Rick Boozer. "They were not learned through any conscious effort on my part; that is, I sort of assimilated them through osmosis. How about the theme to *The Lawman* or *both* verses of *Wyatt Earp?* Want to know the lyrics to the Dr. Caldwell's laxative commercial?"

"The early commercials were very interesting," Marilyn Chadwick exclaims, adding:

> For instance, the host of the show often incorporated the commercial into the program, as did George Burns, who might mention how great Carnation milk "from contented cows" tasted while Gracie was serving him a cup of coffee as part of the skit.
>
> Often a commercial consisted of simply holding up the product and telling the audience what it was. Frequently the item appeared on next week's grocery list at our house, said list also including precise detail as to size and color of box, etc. I remember Mom's exacting speeches to Dad instructing him to buy no substitutes or alternative brands!
>
> Still another commercial that had us sold was for Tame hair rinse, which claimed, "no more tangles" when the comb was drawn through our freshly washed hair. This proved to be true, and every Sunday night we followed our Lustre-Creme shampooing with Tame rinse.
>
> In one instance, I remember the sponsor being visible throughout the entire show. Remember the quiz show *Strike It Rich*, with Herb Shriner as host? His podium had the Old Gold logo on it, and the two dancing packages of cigarettes performed on every show. The girls inside those props were concealed except for their legs and tap-dancing feet. This explains why my sister and I were often seen with our taps on our toes and boxes over our heads!

In the early days, anything appearing on television might develop a following, as Linda Law attests: "Mostly, everyone loved whatever was on. The medium was so fascinating that no one turned his or her nose up at anything. Even the commercials were popular: the dancing Old Gold cigarette boxes, the home appliances, cars —'See the USA in your Chevrolet'— peanut butter—'If you like peanuts, you'll LIKE Skippy!' I always wanted Art Baker to say ' . . . you'll LOVE Skippy' because it had more punch to it." Ruth Watkins speaks for many parents when she says, "The commercials caught the ears and imaginations of my daughters, because they could say and/or sing most of them word for word during those years. The Pillsbury Doughboy (used sometimes to refer to someone we knew), Oscar Meyer wieners' cheerful song, Life cereal (with the older brothers pushing the bowl of cereal toward the younger one and discovering 'Look! Mikey likes it!'), Shake & Bake (with the little girl at the end adding ' . . . and I helped,' which became a line we used often) were just a few of them." Elizabeth M. Blakely remembers, "Very few commercials had an impact on us in terms of purchases, but I recall enjoying the Coke and Cheerios commercials and once convincing my mom that I would eat a bigger breakfast if it was Lucky Charms instead of Shredded Wheat."

R. L. Vaughn will never forget one "live" commercial that he saw in 1952 or 1953 while he was stationed at Lockbourne Air Force Base in Ohio: "While I was having dinner with my wife and children, we, like everyone else, had the newly acquired TV turned on. During a commercial that was demonstrating how to use a certain washing powder, a beautiful, *well endowed* young lady was bending over an old fashioned rub board, and as she bent over, both breasts fell out in full view. The next day at work that was the talk of the entire base, everyone asking each other, 'Did you see that commercial last night?' And needless to say everyone was glued to the TV the next night and, of course, it didn't happen again."

One commercial looms more innocently in Skip Eisiminger's memory:

In the early '50s my youngest sister and I watched *Howdy Doody* on the "flickering blue parent" every chance our mother gave us in the basement recreation room of our house in Falls Church, Virginia. We were mesmerized by the adventures of Buffalo Bob, Clarabelle, Howdy, and the gang in the Peanut Gallery. Best of all, however, was the cannon. One of the show's sponsors, Quaker Puffed Rice and Quaker Puffed Oats, had a big black cannon that we loved, but which some pacifistic Quakers took offense at, I later learned. At the end of each commercial, a charge of oats or rice was rammed down the mouth of the cannon, only to explode into the camera seconds later. We watched in great anticipation, perched on our father's overstuffed red-leather reading chair. When the howitzer was discharged in a flurry of fluffy grain ("*shot* from cannons!" was our cue), we would feign a mortal wound and roll off the arms of the chair onto the carpet or tumble from the chair's back harmlessly into the cushioned seat. I never enjoyed my sister's company more than when we convulsed together in laughter, rehearsing our mortality most every weekday afternoon.

Notes

1. The loop mentioned here and discussion about its "live" authenticity may be seen at http://www.richsamuels.com/nbcmm/wmaq/x9xapnewsreel.html.

2. Alex McNeil, *Total Television*, 4th ed. (New York: Penguin, 1996), 59.

3. Enid Nemy, "Metropolitan Diary," *The New York Times* (February 7, 1999): 45.

4. "Daytime TV Won't Cure Flu, But It Helps," *Atlanta Journal-Constitution* (February 7, 2000): sec. E.

5. Contributed item, Joe Rogers, ed., "Metropolitan Diary," *New York Times* (August 25, 2003): sec. A.

6. Contributed item, Joe Rogers, ed., "Metropolitan Diary," *New York Times* (May 24, 2004): sec. A.

7. "Accidental Learning," *Herald-Leader* (July 11, 2001): sec. B.

8. Butler, "Accidental Learning," *Herald-Leader* (July 11, 2001): sec. B.

9. Amy Blackmarr, *Going to Ground: Simple Life on a Georgia Pond* (New York: Penguin, 1998), 90–91.

10. David Benjamin, *The Life and Times of the Last Kid Picked* (New York: Random House, 2003), 48.

11. "See Some Good Stories Saturday," *Greenville News* (August 14, 2003): sec. B.

— 6 —

News Programs and Events Coverage

"And NOW . . . [pause; then snappily] the NEWS," radio announcers used to say pretentiously and portentously. But present-day television news—a briskly choreographed medley of changing visuals, satellite interviews, and moving color images from nearly anywhere on Earth and some places beyond—was a long time coming. The painfully slow, tentative start of TV news in the late 1940s can be attributed not only to the technical difficulties of synchronizing limited visual material with studio voices but also to the reluctance of network radio newsmen to give up their established ways of relaying events.

Television itself made news on January 11, 1949, when the East-to-Midwest coaxial cable was put into service, and President Harry S Truman's opening address of the Japanese Peace Treaty Conference in San Francisco on September 4, 1951, covered live by all four networks, inaugurated coast-to-coast telecasting. One of Dick Eyman's television firsts was seeing, a few weeks later, "the live interconnection of networks from coast-to-coast. This was dramatized by showing a live picture of the George Washington Bridge followed by the Golden Gate Bridge on a split screen." It was, in fact, the first *commercial* live coast-to-coast telecast, the November 18, 1951 premiere of *See It Now*, a video adaptation of Edward R. Murrow and Fred W. Friendly's CBS Radio series *Hear It Now.*

Harry Truman's was the first televised Presidential inauguration, also carried by all four then-existing networks on January 20, 1949. Mary Lynn Moon, who belonged to an early TV-owning family, notes, "My mother invited the sixth grade from my elementary school to watch. I was 8, so I remember all the big kids' coming to our house as a special day." Dick Stanley, attending a fortieth anniversary reunion for the high school class of 1961, was reminded by a classmate that "when Eisenhower was inaugurated in 1953, there were only three members of our fourth grade class whose families had televisions. The class was divided up among the three of us, and eight or ten came to our house to watch the inauguration 'live.' This was a vivid memory for him, and yet I had completely forgotten it."

While East Coast anchors and producers concentrated on tweaking the quarter- and later half-hour evening news formats in the early television years, a poignant event in 1949 did much to establish the West Coast penchant for sustained coverage of earthquakes, freeway chases, and other protracted happenings. On April 9, four-year-old Kathy Fiscus fell down a well while playing in her San Marino, California, neighborhood, and across the nation the rescue efforts became the lead story in newspapers and on regularly scheduled radio and television newscasts, supplemented toward the end by dramatic bulletins. However, Los Angeles reporter Stan Chambers stayed on the air to provide twenty-seven and a half hours of continuous coverage for KTLA until the girl's body was brought to the surface. Later he said, "We were so wrapped up in the tragedy and reporting the story that we didn't really have a chance to wonder if anyone was watching; we had no idea of the impact we were making." The station describes its coverage of the rescue attempt as "the world's first extended news broadcast," and Stan Chambers alone, according to KTLA's Web site, has since covered more than 20,000 events, including the Robert Kennedy assassination and the Rodney King beating and subsequent riots.[1] Jim Bivens, who had seen Chambers' reporting of the Fiscus story on his self-assembled kit television set, enjoyed an exchange of letters with the TV personality in 1987 and learned that KTLA counted 250 to 350 set owners on the station's mailing list at the time of the episode. By extension, Jim Bivens figures, the total number of sets in Los Angeles area homes had been well under 1,000 in 1949, but "literally thousands of people watched the rescue [efforts] on TV, gathering in homes where the lucky few sets existed, or on corners outside TV stores, or in bars that had TV. [This] was the event that convinced many people to buy televisions."

In the 1950s those TV sets provided grim Cold War-era images of the Army-McCarthy hearings in Washington (meant to expose communist infiltration and influence in many areas of American life), war in Korea, and political infighting at the major parties' presidential nominating conventions. Roger Rollin notes that the Korean "police action" was not a "television war" in the way that the Vietnam conflict would be. Rather, television provided limited images of the Korean fighting similar to the newsreel footage shown in movie theaters at that time.

"In 1954," says Mike Newton, "it was a very hot summer here in Ohio. My mother was ironing clothes in our bedroom, watching the Army-McCarthy hearings. She detested McCarthy and would get so incensed about how he was controlling the hearings. She loved Joseph Welch, the judge who later appeared in *Anatomy of a Murder* as the presiding judge. I didn't understand why McCarthy was able to talk out of turn after the judge told him to be quiet." Mary Ann McKenzie, noting that her mother was task-centered and wasted no time on daytime television's usual fare, found a different situation during the Army-McCarthy hearings telecasts: "Every single day that I came home from school there would my mother be with her ironing board, ironing in front of this thing, absolutely fascinated. I had a teacher at this particular time who thought McCarthy was wonderful. My mother did not think McCarthy was wonderful. My mother thought he was terrible. My teacher shouldn't have done it, but she would talk about how this country was being saved by McCarthy. And I would just sit there and keep my mouth shut, of course. Then I would go home and there was my mother, terribly upset by this man." Adult viewers of the U.S. House Committee on Un-American Activities hearings grew increasingly alarmed at the bullying style of the senator from Missouri (the committee's chairman), and most were gratified when Edward R. Murrow took him to task in a series of *See It Now* programs, culminating in the broadcast of March 9, 1954, when Senator McCarthy grudgingly accepted Murrow and Friendly's invitation to respond on film to the series' recent characterizations of his interrogation tactics. Murrow ended the program with a defense of his own journalistic purpose. In later hearings McCarthy began to rail against his perceived attackers, including Murrow and Army Counsel Welch, whose own response "Have you no decency, sir?" crystallized growing public disgust with McCarthy's televised hearings and prompted the Congress to censure the senator before the end of the year. Seasoned broadcast commentator Elmer Davis commented later that the McCarthy-Murrow confrontations "taught everyone who works in television that the medium can show up a man's character and his record . . . by putting on the screen his own actions and words."[2]

Television showed—and provoked—tensions during the 1950s Democratic and Republican nominating conventions too. Mary Ann McKenzie confesses to peeking over the stair rail to watch the political parade when the 1952 sessions ran beyond her designated bedtime. She was especially drawn to Tennessee's Senator Estes Kefauver, who led the first two ballots in the Democratic presidential nomination but had to settle for the vice-presidential slot on the ticket with Illinois Governor Adlai Stevenson: "I remember being fascinated by Kefauver as a personage—you know, just gape-mouthed looking at this man on the television. He was so dramatic. It's funny what fascinates children." On the other hand, her self-important uncle was not pleased by television or the progress of the 1952 Republican convention:

He was a judge, and at this time he was working for the Justice Department. They had no television, because they were too good for television. But they were visiting that summer when the conventions were on. And he was entranced. Now, he was this extremely dignified individual. I didn't like him much, actually, as a child because he wasn't very child-friendly. He'd walk into a room and you'd want to stand up just because he walked in. He was this big, imposing fellow, and terminally cranky. Anyway, I never saw him get animated about anything until I saw him sitting down with my family watching this convention. He was a Robert Taft man, and when they nominated Eisenhower instead of Robert Taft, he about had a conniption fit. He was shaking his fists at the television, and I remember being amazed that something had gotten to my uncle like that. He thought that there was a great injustice being done, which if you were a conservative Republican, it *was* a great injustice being done.

(Laughs). But my father was a Democrat and my mother was a moderate Republican, so they weren't all in an uproar as he was.

Tennessee-born Dixie Ann Schmittou and her family watched the 1952 nominating conventions with special interest because their state's U.S. senator was strongly in the running, building on a reputation gained by his handling of the Kefauver Crime Committee hearings, which had been televised in 1950 and 1951. However, she says, "Adlai Stevenson won the nomination, and I remember the pictures of his shoe with the hole in the sole, which Democrats used to show how humble they were. In 1952, our school had mock elections and I spoke for Stevenson while my best friend spoke for Eisenhower. (Stevenson would be the last Democrat I supported, much to the dismay of my yellow-dog Democrat father.")

On the other hand, Roger Rollin, whose family purchased a television set specifically for the purpose of following the 1960 Democratic gathering, was happy with the results:

The Democratic National Convention was all that I had hoped it would be. I recall being out of the room when Senator John F. Kennedy first addressed the delegates and thinking, a bit disappointedly, "Why he sounds like Jimmy Cagney!" I got over it. JFK's thousand-candlepower charisma, which later that year I was to experience in person at a rally in Lancaster, came across even on a small black and white TV screen.

And, yes, watching those first-ever presidential debates, I thought Nixon looked like a thug, JFK like a movie star. (Yet I admit that, after watching Nixon's televised "Checkers" speech to save his vice-presidential candidacy back in 1952, I was enough taken in to conclude that Eisenhower should keep him on the ticket.)

When JFK picked Lyndon Johnson for the Vice-Presidential slot, it should have been a signal to me that there was a deep vein of cynicism and calculation behind the lofty and idealistic Kennedy rhetoric. But I was as seduced by Camelot on TV as a lot of others. There was Kennedy, bare-headed, at his inauguration, with Jackie by his side, the only truly beautiful First Lady I was aware of. There were the televised presidential press conferences, where the Kennedy wit and humor, intelligence and agility, were on full display. There were the televised White House concerts, the photo ops of Kennedys at play with their photogenic children, and one of the events of the early '60s, Jackie's televised tour of the White House. The Kennedys were *made* for television as the Nixons (in spite of his having an attractive wife and two good-looking daughters) were not. Unlike the Kennedys, none of the Nixons ever appeared comfortable on television.

After the 1941 Japanese attack on Pearl Harbor, nearly everyone in the country remembered where and when he or she heard the radio bulletin of the news, and the event is often cited as a turning point in the history of radio news. Arthur Godfrey's description of President Franklin Delano Roosevelt's funeral procession for CBS listeners on April 14, 1945, was another of those singular radio moments.

Television news's coming of age is commonly dated at the four days beginning on November 22, 1963, when President John F. Kennedy was fatally wounded by shots fired from a schoolbook depository into his Dallas motorcade and ending with his burial at Arlington on the following Monday. A "CBS Special Report" graphic interrupted the Friday episode of *As the World Turns*, and Walter Cronkite made the initial audio-only announcement that President Kennedy was reported to be "seriously wounded" after three shots had been fired in downtown Dallas. Soon Cronkite appeared in shirtsleeves to preside over the live field reports that began to filter in. At 3:27 P.M. Eastern time he was handed an Associated Press wire report. He paused for a moment and then spoke in shards of phrasing: "From Dallas, Texas, the flash, apparently official. President Kennedy died at 1:00 P.M., Central Standard Time, . . . 2:00, Eastern Standard Time." In a visual "gulp" seen across the nation, Cronkite, his head turning slightly as he fought back tears, lifted his thick-framed eyeglasses off his nose in an unconscious gesture and then resumed presiding over a long on-camera shift as further details filtered in and cameras were put into place at Parkland Hospital, where Kennedy had died. On ABC-TV about two hours later, anchor Frank Reynolds, frustrated by uncertain information and feeling the general shock of the event, seemed to lose control momentarily. Turning to the newsroom personnel just off-camera, he said in a near-shout, "Listen! Let's get it right; let's . . . get this thing *right*." By that time, all stations had gone to continuous coverage. The weekend program listings of *TV Guide* had been voided by the events in Dallas, and viewers' only choices lay in which channel to trust to convey the evolving story, monitored around the clock without commercials and without station breaks. The four days of uninterrupted coverage proved oddly disorienting to viewers accustomed to their stations' signing off after the late movie, yet many could scarcely turn away from it. Kathy Chastine-Burrell and her husband replaced their 1955 TV set with a new one "to watch all the things about that."

Roger Rollin provides one man's recollection of television's sustained coverage on and beyond the assassination day:

I was teaching an afternoon class . . . when two students I had never seen before walked into my classroom, their faces strained and serious. "We thought you would want to know, the President's been shot in Dallas." I dismissed the class, got into my car, turned on the radio, and headed home. The news came: JFK was dead. My wife had not heard of this national tragedy, and we turned on the TV to learn more. Gradually, television sorted things out for us. The motorcade. The shots in Dealy Plaza. The rumors of multiple assassins. The capture of Lee Harvey Oswald. The casket being loaded into the plane and Johnson taking the Oath of Office with Jackie, still in shock, at his side. We seldom had the TV off, and when, as I watched, Lee Harvey Oswald was himself shot and killed, on camera, I was in shock. What country are we living in? And that question arose again when television showed the balcony of a Memphis motel on which Dr. Martin Luther King sprawled, shot to death. And when we saw footage of Bobby Kennedy, now a presidential candidate, giving a triumphant speech in a Los Angeles hotel and then, moments later, lying in his blood on the floor of that hotel's basement. Television seemed made for tragedy. (See the Challenger disaster. And 9/11.)

The around-the-clock television reporting of the John F. Kennedy assassination affected even the very young. The familiar cartoons and sing-along commercials had disappeared, adults were preoccupied, and usual routines were suspended. Vickie Metz says, "I was five years old the day President Kennedy was shot. Over the next few days I watched the unfolding drama with my parents and grandparents. It was the first time I had ever seen grownups cry. I realized for the first time that the world was not always a safe and happy place."

The Monday funeral day, designated by the new President as a National Day of Mourning, brought further indelible pictures to the television audience. After the funeral mass at St. Matthews Cathedral, the symbolic riderless horse Black Jack, kicking up, seemed at first to be agitated by the events but soon adapted to the ceremonial solemnity of the occasion. As the caisson and the honor guard passed in front of the church, the President's son "John-John" gave a brisk salute on what the commentators noted was by sad coincidence the child's third birthday.

Beth Jarrard regards the John F. Kennedy assassination as "a defining moment in my life," explaining,

> Along with millions of people around the world, I watched almost every minute of the Friday-Monday broadcasts . . . . I couldn't believe my President—the one I had voted for in my grade-school mock election—had been assassinated. I missed the shooting of Lee Harvey Oswald by Jack Ruby. Our minister had preached a little too long that Sunday. I guess he felt we needed it after the terrible thing that had happened, but we arrived home and turned on the TV just in time to see the commotion after the shot was fired. I had often heard my parents talk of seeing the funeral train bearing Franklin Roosevelt's body come through Greenville. I wasn't in Washington to see the [Kennedy] funeral procession or the burial, but I SAW it all on TV, and it is an image that will never leave my memory.

"Everyone remembers when Kennedy was shot," sums up Dixie Ann Schmittou; "I remember holding my two-year-old son and crying because John-John no longer had a father. That next Sunday, I was watching when Lee Harvey Oswald was shot—live on TV." Thomas Doherty of the American Studies program at Brandeis University, who has written extensively about the broadcast coverage of those events, calls the network efforts "the most moving and historic passage in broadcast history" and describes "the quiet power of the spectacle [as] a masterpiece of televisual choreography." He further asserts about those four days in 1963, "The saturation coverage . . . yielded a shared media experience of astonishing unanimity and unmatched impact, an imbedded cultural memory that as years passed seemed to comprise a collective consciousness for a generation."[3] CBS news producer Don Hewitt, creator of *60 Minutes*, says, "Walter Cronkite went on the air and stayed there for 36 hours. Americans that morning did not go to church. They went to their television sets. Walter Cronkite single-handedly calmed this country down after the assassination."[4]

"I couldn't keep my eyes off the television," Mary Ann McKenzie says of the time of the Robert Kennedy and Martin Luther King assassinations. She continues,

> It was not as bad as the shock over [John F.] Kennedy, but it was shocking. I was teaching at the time at Blacksburg High School right next to Virginia Tech in Blacksburg, Virginia, where I got my master's degree. I had a group of black students, and I went to school the next day after Martin Luther King was assassinated, and they all wanted to talk to me. And I was frightened because these kids who had seen it at home on television, it was like somebody had hit them. I can't tell you how much pain they were in. But they had come to school, interestingly enough, because their mother and father had said, "You get up and you go to school." But they were in mourning and they were ready to fight somebody, and I just talked to them until they calmed down, and they got through the day and I got through the day.... That whole period just shook me up. And the assassination of Robert Kennedy took place just about the same week I found out I was pregnant with my daughter. So, I just felt like the whole world was upside down.

With one child born in February 1969 and another in August 1970, Mary Ann McKenzie found her viewing habits changing: "When I had to nurse the baby at five in the morning, I'd be watching *Farm Today* and all sorts of stupid things, but not the news. I've always been someone who watched the news at night, but that was a time when I didn't watch it so much because it was so . . . depressing."

President Kennedy had sent a limited number of military advisors to Vietnam, but his successor, Lyndon Johnson, led the country into full-fledged involvement and saw public opinion sharply divided on the need and the conduct of American participation. Dixie Ann Schmittou recalls the growing unease of the day: "We would see the war in Vietnam on the evening news. My youngest son was born in Savannah, Georgia, in 1966, and we had good neighbors who had a ten-year-old son. I remember him anxiously asking his parents about having to go to war in a few years. At that time, we weren't sure ourselves if it would be over before he would be called." To answer such growing public concerns, Walter Cronkite took a CBS News crew to Vietnam during the particularly bloody Tet Offensive of early 1968 and sent back a series of grim reports. On his return home, the newscaster who was often called "the most trusted man in America" ended the February 27, 1968 edition of the *CBS Evening News* with an editorial segment that saw little chance of the Vietcong forces being defeated anytime soon: "To say that we are mired in stalemate seems the only realistic, yet unsatisfactory, conclusion." At the White House, President Johnson came to an equally grim conclusion about the fate of his presidency, confiding to his associates, "If I've lost Walter Cronkite, I've lost middle America," and not long afterward he announced that he would neither seek nor accept his party's nomination for reelection. "I remember feeling very shocked," Mary Ann McKenzie says of that unexpected conclusion to President Johnson's March 31, 1968 White House television report on the war situation.

The John F. Kennedy, Robert F. Kennedy, and Dr. Martin Luther King assassinations in the 1960s proved to be dominoes falling toward a general gloom in the war- and social discord-beset 1970s. *Life* magazine proclaimed 1971 to be the twenty-fifth anniversary year of television and devoted the second half of its September 10 issue to a largely downspirited celebration of the occasion. Having commissioned Louis Harris and Associates to poll the American public about its TV-watching habits and expectations, the magazine reported that while regular network news programs earned the highest approval among all program categories, "For the programs which make up the bulk of TV fare, the results range from discouraging to disastrous." The poll also appeared to show that "The day when viewers were glued to the set seems to have passed" and that "The heaviest TV viewers of all are those Harris characterizes as 'lonely and alienated' [while] the busiest and most influential among us, the affluent and best educated, still devote approximately two hours a day to television."[5] Against those impressions, Randy Cox interestingly describes his family's varying degrees of regard for the television set in that day:

> My father watched from time to time, but I think he enjoyed reading the newspaper more, and my mother was always around the set, doing something else like crocheting, reading, or carrying on a conversation. It seemed that the TV was expected to act as "white noise" in the background.
>
> To a kid, news telecasts held little excitement. Often, my brother and I didn't come in from playing with the neighborhood children until just before Prime Time. I don't think my mother minded this practice, though it meant we usually gobbled down dinner after everyone else was finished, because this way we weren't exposed to the unseemly side of television. She was, and still is, quite a follower of current events, but I don't believe she wanted her sons watching carnage from Vietnam. In fact, the only news I remember finding of interest at that time was a piece on the traffic jams and unusual happenings of Woodstock. I remember asking what that was all about, and my mother turned the channel to keep us from watching such abhorrent behavior.

While the saturation drugs and rock and roll of that muddy festival on a New York farm held the interest of some TV news viewers, events in the U.S. space program brought extremes of exuberance and pain to home spectators. Young viewers stayed up past usual bedtimes to see the first manned moon landing on July 20, 1969 and to share the enthusiasm of Walter Cronkite, who rubbed his hands together and repeated "Oh, boy . . . oh, boy" in almost childish glee as, at 10:56 Eastern Daylight Time, Neil Armstrong stepped onto the lunar surface and said, "That's one small step for [a] man, one giant leap for mankind." Tom Worsdale remembers it: "I was in Stamford, Connecticut, with my cousins at their house, and we were all watching. It was thrilling, but I have to admit the quality of the transmission left a lot to be desired. The effect was a lot greater in one's mind than on the screen." Later that night, viewers saw President Nixon, linked to the spacecraft from the White House, congratulating astronauts Armstrong and Buzz Aldrin after their two hours of rock sampling and flag planting on the lunar surface. This and other space journeys in that era brought "special report" telecasts of suspenseful splashdowns for returning crews.

Beth Jarrard found personal ties in such televised occurrences:

Some of the most profound and touching moments of TV for me have been news programs and broadcasts of actual events. My father, the TV dealer, often brought a portable model to my elementary school for such events as the 1960 wedding of the British Princess Margaret to Anthony Armstrong-Jones and the launch of the spacecraft carrying John Glenn. I later watched many launches on TV and in the skies over Central Florida, where I lived in the '80s, but none was more moving to me than one in November 1981. I had returned home to South Carolina to spend some time with my father, who was dying of cancer. From his hospital room, we watched the launch of the Space Shuttle Columbia—one of the first—and he cried. I knew it was because he knew he was watching history in the making AND the beginning of an exciting era in our country, and he wanted to live to see it. He died two weeks later.

Against the triumphs of Russian and American space efforts came the NASA Space Shuttle Challenger disaster at Cape Canaveral, Florida, on January 28, 1986. In the late morning schoolchildren watched in their classrooms as Challenger, whose crew included Sharon Christa McAuliffe, the first teacher selected to go into space, exploded in plumes of smoke seventy-three seconds after liftoff. As the network cameras panned the crowd seated in stands at the launch site, TV viewers could see their faces, including those of Mrs. McAuliffe's family, register triumphant smiles, then uncertainty (many thinking at first that the smoke plumes might be merely a picturesque byproduct of the launch), and then grief. Walter Cronkite, by that time widely known to be a great enthusiast for the U.S. space program, was left again with the task of articulating events that seemed cruelly unexpected.

For Linda Howe too, relatively recent TV events were reminders of earlier griefs brought home by television:

The most disturbing news bulletin: Princess Diana's death. I felt like I knew her. But the worst one happened when I was sitting in the living room with my foreign exchange student son, who had romantic dreams of entering the military, when the first news flash of the Gulf War [1990-91] hit the screen. It immediately flashed me back to Vietnam and all the years of horror on TV at suppertime. At the end of the evening, Siggi (the student) no longer thought war was romantic. In 30 minutes he had grown up in front of my tear-filled eyes. That reminded me of all the horrible bulletins early in the Vietnam War, through junior high, high school, and the early years of my marriage. It made me hate war, distrust my government, and cry for people on an individual basis. I couldn't hate a country. They were individuals.

Linda Law, describing herself and her husband as "really picky viewers today," remembers the early-days fifteen-minute newscasts of John Cameron Swayze: "That got me hooked on news very early." Richard Leiby, who also knew the first-generation quarter- and half-hour newscasts, notes a later vivid fusion of TV news and "real life": "I was leaning in the doorway to our kitchen in Garden Grove, California, about 1967 or '68, watching the evening news, when I began to feel a little unsteady, at which moment the newscaster said, 'We're having an earthquake!'" Just such considerations have caused Margie Chapin to keep her TV receiver at hand: "Here I am 46 years later rarely watching TV but not brave enough to not have it in the house in case there is very important news (i.e., the Gulf War) or to watch video movies or in cases of extreme boredom."

Notes

1. http://ktla.trb.com, accessed June 25, 2003.

2. Quoted in Dennis McCann, "Nation Saw McCarthy at His Worst on Murrow's CBS Broadcast," *Milwaukee Journal-Sentinel* (October 1, 1998, rpt.), http://www.jsonline.com/news/wis150/stories/101sesq.stm, accessed June 25, 2003.

3. "Assassination and Funeral of President John F. Kennedy," http://www.museum.tv/archives/etv/K/htmlK/Kennedyjf.htm, accessed June 26, 2003.

4. Quoted in Dana Calvo, "'The President's Been Shot,'" *Smithsonian* (November 2003): 87.

5. "But Do We Like What We Watch?" *Life* (September 10, 1971): 42–43.

$$-\!-\!-\ 7\ -\!-\!-$$

Informational, Cultural, Quiz, and Sports Programs

In television's vintage days, the efforts of network news divisions extended into broadly informational and educational programs. Cultural programs often reflected the tastes of network founders or the status images sought by corporate sponsors. Mindful of the Federal Communications Commission's (FCC) "in the public interest" expectations, commercial networks carried, often on a sustaining basis, types of programs that now appear chiefly on PBS or on subject-specific cable channels. Even early quiz programs emphasized knowledge rather than the modest prizes offered, and sports coverage generally lacked the blockbuster promotion, technical sophistication, and schedule dominance that it enjoys today.

After the Sunday late-morning and midday public affairs forums *Meet the Press* on NBC and *Face the Nation* on CBS, viewers could choose religious programming, *Zoo Parade* from the Lincoln Park Zoo in Chicago, *Wild Kingdom* from reasonably rugged places, and the anthology program *Omnibus*, hosted by Alistair Cooke, a BBC radio correspondent contentedly settled in the United States. The resident drama and music companies of Lincoln Center were frequently on display. Before event-to-event sports programming came to dominate Sunday afternoons and early evenings, viewers were fascinated by *Victory at Sea, Air Power*, and *The Twentieth Century.*

Omnibus began on CBS in 1952, passed to ABC for its 1956–1957 season, and ran on NBC from 1957 to 1959. "I used to watch the original *Omnibus* program," Mary Ann McKenzie says, "and Alistair Cooke would look into the camera so you'd see just his face, you know; it was like he was talking directly to you, and you'd sit in front of the television like this (makes a "transfixed" face). And he would get things from the Metropolitan Museum of Art and say 'Look at this,' and it was like show and tell on television back then, and I was fascinated by that." She especially remembers the November 8, 1953 *Omnibus* presentation of Aaron Copland's "Billy the Kid" ballet, and she appreciated the variety of subjects presented in the first seasons, including reenactments of episodes in the lives of Anne Boleyn and Abraham Lincoln, features on high-speed photography and on photographer Phillippe Halsman, an examination of U.S. place names, and a performance by the Benny Goodman Trio. Leonard Bernstein began his association with *Omnibus* in 1954, and his recreation of Beethoven's composition process for his Fifth Symphony gave a memorable glimpse into the creative mind at work.

"My mother loved *Omnibus*, a sort of precursor to *Nova* and the Discovery Channel," says Linda Law, who herself vividly remembers the effect of the program's opening: "The theme music was from Gustav Holst's *The Planets*. The piece was 'Mars, Bringer of War,' and it played over a scene of some African tribesmen, very tall and thin, and it scared the hell out of me every time! The music, not the tribesmen."

Born in 1949, Giles Singleton was spontaneously drawn to music and dance of all sorts on 1950s television. "I remember dancing to 'Swan Lake' with the ballerinas and seeing people play the violin, harp, and flute. I chose the flute for my instrument in band." Her grandparents bought a black-and-white set a few months before her parents did, and "When my grandparents got their TV, we would go through the vacant lot to their house, almost next door, and watch some shows with them. I remember my grandfather fuming about 'Liber-AT-chee,' though my grandmother, a classical pianist, insisted on watching every time he came on. At two or three, I could not understand why Papa didn't like Liberace. I loved the diamonds and sequins and costumes."

Less than a decade after the end of World War II, NBC's 1952–1953 season offered the twenty-six-episode *Victory at Sea*, with a notable score by Richard Rodgers of Broadway musical fame, and in many families that epic series vividly answered the frequent question "What did you do in the war, Daddy?" "My dad spent WWII in the South Pacific on subs and a sub-tender as a Navy non-com periscope operator," Linda Howe says. *"Victory at Sea* was his very favorite! He did some critiquing, though, from time to time—as both of us did if any German ever appeared on any show—since we spoke German at home. You wouldn't believe what they sometimes said—cracked us up! Non-German-speaking people wouldn't even know—I'm sure the German speakers of the U.S.A. were rolling sometimes, though. And it wasn't their horizontal or vertical controls out of whack!" Art Donahoe's family had a separate rationale for each program in an extensive smorgasbord of shows that it watched. For instance, he says, "With WWII fresh in the minds of my parents, we watched *Victory at Sea,* [just as] being a good Irish-Italian Catholic family, we watched Bishop Fulton Sheen."

Mary Ann McKenzie would watch *Victory at Sea* "religiously" by herself and *Person-to-Person* and *See It Now* with her family: "And that may be why I was so good at history, because, at an impressionable age, I was watching every one of those programs. In fact, WWII was very important in my neighborhood. The neighbors on one side of us were from London, and they had immigrated to America because they were afraid London was going to be overrun during the bombing. On this side, we had a Jewish family, and they had relatives who had been through the Holocaust, and when I would visit them they would tell me stories of the Holocaust. When I run into people my age for whom the Second World War was not a big thing, I'm always amazed, because for me it was really important."

In 1998, Pulitzer Prize-winning historian Doris Kearns Goodwin told interviewer Leah K. Glasheen that she is intrigued by the transforming role of television, a relatively new technology in her youth. In its earliest days, Goodwin says, television had a much different social role, often blurring fantasy and reality in a manner inconceivable to today's more sophisticated viewers. She remembers vividly the day she looked at the television set and saw Joan of Arc being burned at the stake. "There, before my eyes, the young woman stood, lashed to a piling atop a pyre," Goodwin says. Horrified, she ran to her mother, who explained she had simply seen a dramatization. Afterward, Goodwin eagerly absorbed all of the *You Are There* series narrated by Walter Cronkite (the capture of John Wilkes Booth, the siege of the Alamo, the fall of Fort Sumter, and so on). TV's role as "perhaps the most fragmenting force in all our cultural history," Goodwin says, "is so ironic [especially since] back in the 1950s it created more bonds among us rather than [fewer]."[1]

Another illustration of Doris Kearns Goodwin's suggestion that 1950s TV blurred "fantasy and reality" is what Leah Glasheen calls "a vicious sort of soap opera that had American families glued to their new TV sets: the McCarthy era. . . . Goodwin tells of how she and her friends concocted their own version of McCarthyism, played out in living rooms throughout the neighborhood. "The person designated as the accused sat in the chair while the rest of us asked questions and made charges from behind the table," she says, "with the charges becoming more mean-spirited as the game went on."[2]

Al Raymond was more quietly fascinated by the syndicated program *Industry on Parade*, produced by the National Association of Manufacturers between 1950 and 1958, although he is not certain that it had much to do with shaping his career in design and engineering. Nonetheless he found this program "kind of neat" because "They would show you different manufacturing things; they would take you to different plants and show you how different things were made, how different things were done. It was fascinating to see all these little gadgets and machinery they showed." For such elucidations *Industry on Parade* won a Peabody Award in 1954.

"Infotainment" is a relatively new term used to describe shows, like those on the Animal Planet, Oxygen, or Home and Garden channel today, which can inform without seeming to lecture the viewer. Here such "heavy" elements as specific dates and geographical locations, Latin names of flora and fauna, and other "academic matter" are minimized or omitted, and the prevailing tone is breezily friendly. In differing ways, some programs from (or begun in) the early television years accomplished the same balance of informing and entertaining. Falling broadly into this category are quiz programs, sports shows, and Sylvester "Pat" Weaver's *Today* and its imitators.

"The quiz shows were, of course, so interesting that no one has improved on the earliest format," says James Andreas, Sr., adding, "These shows got me interested in knowing things, learning about new ideas. So we still have *Jeopardy*." In these prime time programs, panelists from the New York entertainment and journalism worlds tried to get to the bottom of two guests' impersonations of a third (*To Tell the Truth*) or to pin down the occupation (*What's My Line?*) or a distinguishing achievement (*I've Got a Secret*) of a "mystery guest." The opening scene of Steven Spielberg's 2002 film *Catch Me If You Can*, based on the autobiography of famed impostor Frank Abagnale, Jr., recreates a 1977 episode of *To Tell the Truth* under its third host, Joe Garagiola, who has observed that such shows "are not stupid, they're enjoyable" and that in watching them " . . . you learned without knowing you were learning, which was good."[3] Music critic Terry Teachout has recalled that he was allowed to watch *What's My Line?* late on Sunday evenings if no "high crimes or misdemeanors" had recently marred his behavioral record, and after more recently seeing the program again on tape, he observes, " . . . the collegial bonhomie of the participants leaves you with the distinct impression that the show is taking place in a parallel universe of famous people who all know and like one another and probably stroll over to the Algonquin for a drink afterward. Or so, at least, it seemed to me when young, sitting in front of a black-and-white TV in the living room of a small house in a small town in southeast Missouri."[4] In Marilyn Chadwick's childhood home the late time slot of *What's My Line?* was a problem: "As much as I loved seeing it, the experience was bittersweet, because it was always associated with having to have all my homework caught up, so I'd be ready for another Monday morning at school. The panel consisted of Dorothy Kilgallen, Arlene Francis, Bennett Cerf, and Fred Allen."

Daytime network and later syndicated quiz and other audience participation shows were less sophisticated. Arlene Francis's quietly bemused smile as she plotted an incisive question was replaced by the screech of the studio audience urging on a favorite contestant. Marilyn Chadwick says that her family "watched scores of game shows, all of which were entertaining and challenging," and she continues,

> There was *To Tell the Truth, Password, Jeopardy*, and the original *Hollywood Squares*, with Wally Cox, Paul Lynde, Charlie Weaver [played by Cliff Arquette], Rose Marie, Zsa Zsa [Gabor], and Gypsy Rose Lee. When my own children were very young, we loved to tune in *Hollywood Squares* every morning.
>
> *Beat the Clock* was really fun. Contestants would race against time, doing all kinds of laughable assignments. Mom watched this one very seriously, paying a great deal of attention to detail. She would recreate the little obstacle course and props in our basement, and we would invite all our neighborhood friends over to play. Once we had a toupee swinging from the overhead floor joists on a string, which we had to land inside a top hat on our head. I watched Mom give an explanation of the swinging toupee to a very uneasy-looking meter man one day.

No special display of intellect was required for participation in *Queen for a Day*, which came to the NBC-TV daytime schedule in 1956 after more than a decade on the Mutual Radio Network. Each day host Jack Bailey, speaking with hearty enthusiasm, invited three women from the audience to tell of a need in their lives—a Cinderella wish that would be granted to the contestant whose story most affected the audience and thus registered most strongly on the "applause meter" near the end of the program. Other gifts, typically including a set of luggage, a beauty treatment, and a limousine tour of Hollywood, completed the daily prize package. Cynical at-home viewers, not to mention the more cunning contestants, realized that some requests were almost certain winners: a plea for a Bible, a need expressed on behalf of a helpless child, or a story involving the poignant wish of a desperately ill or dying mother. On any day when one of these classic appeals clashed with another, the voting period could be especially tense. Speaking for herself and her sister Priscilla Kanet, painter Caroline Jasper says, "It's hard to believe that we (Mom included) were so into *Queen for a Day*. Each contestant tried to out-sob-story the rest. I wonder if we felt sorry enough for the ones who didn't get picked to be Queen. We were too busy oohing and ahhing over the robe and crown [the winner] got to flaunt like Miss America. What a value system! No wonder so many Baby Boomers' first marriages hit the skids."

In the first generation of quiz shows, cash prizes were very modest, often in the two- or low three-figure range, with a bonus supply of the sponsor's products: Max Factor beauty aids or a carton of Camel cigarettes. With a second generation of quizzes the winnings jumped much higher, as did the viewership levels. Thus the 1940–1952 radio quiz *Take It or Leave It* offered prizes beginning at $1 and doubling up to $64 as the questions grew increasingly difficult. Retitled *The $64 Question* in its last two radio years, that program became the basis for CBS-TV's *The $64,000 Question*, which debuted on June 7, 1955, with Hal March as host and Revlon cosmetics as sponsor. Drawing huge audiences in its early weeks, that contest show led to competing ones such as NBC's *Twenty-One*, first aired in September 1956. The innocent Eden of quiz shows was spoiled when ratings-hungry producers and sponsors noticed something in the demographics. Roger Ebert, in his *Chicago Sun-Times* review of the 1994 film *Quiz Show*, directed by Robert Redford, explains how the serpent crept into the picture:

A milestone in the decline of American values came in the mid-1950s, when it was revealed that many of the top TV quiz shows were rigged—that contestants were being supplied with the answers. . . .

The early quiz shows rewarded knowledge, and made celebrities out of people who knew a lot of things and could remember them. The post-fix quiz shows rewarded luck. On *The $64,000 Question* and *Twenty-One* you could see people getting rich because they were smart. Today people on TV make money by playing games a clever child can master. . . .

Quiz Show remembers [the 1950s as] a decade when intellectuals were respected, when a man could be famous because he was a poet and a teacher, when TV audiences actually watched shows on which experts answered questions about Shakespeare and Dickens, science and history. All of that is gone now.

The first show, CBS's *The $64,000 Question*, was apparently on the level. But across the street at NBC's *Twenty-One* executives and sponsors watched the ratings, and realized that some contestants drew more viewers than others. A grating know-it-all named Herbert Stempel won for weeks on *Twenty-One* because he was being given the answers. The executives decided his appeal was wearing thin. So they broke the news to him: He'd had a free ride long enough, and now it was time to lose.

Stempel took that news very badly. Meanwhile, America liked his successor, an attractive, disarming intellectual named Charles Van Doren, who was a member of one of America's great literary families.... Blinded not so much by money as by fame, Charles had agreed to cheat. And when Stempel blew the whistle on the whole setup, a Congressional investigator brought the deception tumbling down.[5]

Before the cancellations of all such shows in the fall of 1958, Marilyn Chadwick's family had chosen one contestant to cheer on at home: "Our family was among the first to see Dr. Joyce Brothers launch her celebrity career before a television audience. We followed her as a contestant on *The $64,000 Question*. We were really intrigued by the sound-proof booth, a replica of which Mom was unable to muster."

In sports programming, too, television had had an innocent age before innovators such as Roone Arledge worked their magic with new camera techniques, shrewd pacing, and audacious scheduling. ("Football on TV on a *weeknight?*" viewers exclaimed when *ABC Monday Night Football* was introduced in 1970.) In the early days, time-filling roller derbies could be shown cheaply from local arenas, and for audiences accustomed to watching test patterns patiently, the experience of seeing skaters circle around rinks was entertaining enough. Albert Holt remembers Gillette Razor-sponsored boxing matches featuring Archie Moore, "Kid" Gavlin, Cassius Clay (later known as Muhammad Ali), "Slappy" Rosenblum, and "Rocky" Marciano, and he recalls driving to another town with a group of friends because the television set in a furniture store window there was "rumored to have the clearest TV reception in order to see Notre Dame play the University of North Carolina" in football. Dennis Thompson remembers, "The World Series used to be played during the day, and while many clandestine radios were at school (sometimes a teacher could even be talked into letting the class listen for an inning or two), the really lucky students got to stay home for a day and watch on television!" Wrestling, pronounced "wrasslin,'" brought a cast of colorful characters to the home screen, men whose professional names (such as "Gorgeous George") seemed to be *all* nicknames and whose performances seemed to be guided more nearly by circus tradition than by a sports rulebook.

Former President Gerald Ford, prone to slipping on ski slopes and getting around on crutches, was also liable to slips of the tongue: "I'm a great fan of baseball. I watch a lot of games on the radio, er, I mean television." For James W. "Sam" Sloan, the first glimpse of a televised baseball game proved a very mixed experience:

In the fall of 1951, I was a 14-year-old lad growing up in the small town of Walhalla, South Carolina. I had barely heard of the magic box of TV and had never seen one, even in a store window. I was a die-hard Brooklyn Dodgers fan and had been as long as I can remember. I read every book I could find on their history and listened to as many games broadcast on the radio as I could. Baseball was king in the summertime back then; every textile mill had a team, and on Wednesday and Saturday afternoons the whole town was at the game, it seemed like.

Because I was pretty good at sports, I had made friends with the more affluent kids in town, and on this particular fall afternoon I was going by my buddy Robert Clark's house, and we would walk to midget football practice together. The Dodgers and the New York Giants had tied for the National League pennant and were in a three-game playoff to decide the 1951 league winner. I was listening on the radio and could not bear to leave before the game was over because they had won one game each, and this game was to decide it all, but I knew Coach would get mad if I was late to practice and would probably bench me. So I left to go by to get Robert. When I got there, he asked me if I had ever seen a magic box called television. I said, "No," and he told me his dad had just gotten one and the Dodgers and Giants game was on right now.

Suddenly football practice was not as important anymore. We went into his house, and I could hardly believe, even through a somewhat snowy picture, there were my beloved Dodgers in plain view. Ironically, it was the bottom of the ninth inning, with the Dodgers leading, I believe, by two runs, and the Giants had two outs with two men on base when the Dodgers manager decided to bring in relief pitcher Ralph Branca to face Bobby Thompson. And as history has retold us hundreds of times, Bobby Thompson hit "the homerun heard around the world" and broke my heart on the day of my first experience seeing the magic box of television.

"We got more than the rest of the country," Tom Worsdale says of the sports programming available to New York City viewers. He elaborates,

I can barely remember watching the Brooklyn Dodgers at Ebbets Field during their last season in 1957, but I do [well] remember the Mets as they took to the field for the first time in 1962. I became a lifelong Mets fan. I was also big on the Knicks and the Jets too, and never missed any New York teams when they were on TV. I can remember sitting in front of the TV during both the '69 World Series and the '70 NBA Championships keeping score as the games went along. I made up my own official score sheets and kept track as if I were the official scorer. In fact, I still have some of those sheets stashed away in a trunk to this day. I can also remember tuning in to one of those in-between channels, like Channel 6, which we didn't get in New York, and barely being able to pick up a Philadelphia station, watching it through the fuzz. It was a big deal to get something—a game or a special movie from another city, even if you could hardly see or hear anything.

Television forever changed football (the commercial hype of the Super Bowl being exhibit A), and televised football changed a relative of Caroline W. Todd: "After we got television pictures, my Aunt Francenia Brennen fell in love with football. Her favorite team was the Green Bay Packers. When they were on TV, she would lie on the sofa across from the TV set, put an ashtray on her chest, and smoke at least two packs of cigarettes per game. She never inhaled, but the stress of the games made her puff away on those cigarettes."

Marilyn Chadwick, not counting herself a television sports fan, nonetheless found "a particular technical advancement to be the hallmark in my years of experience as a viewer: THE INSTANT REPLAY. I marveled when Dad first explained and showed it to me. It was a baseball game, and I knew I was witnessing something with the impact to make a real difference in the future of televised sports. Few today can imagine what it was like watching the World Series before the instant replay days."

While Roone Arledge was often called the "resident genius" at ABC-TV because of his innovations in that network's news and sports telecasts, Sylvester "Pat" Weaver made a profound difference at both ends of the traditional TV "day" at NBC. *The Tonight Show*, initially called simply *Tonight!*, was hosted by easygoing Steve Allen in its September 27, 1954 debut, and its further hosts were volatile Jack Paar, laid-back Johnny Carson, and the more stinging Jay Leno. Jack Paar held serious discussions with frequent guest Alexander King and with announcer Hugh Downs, and political figures have taken their turns in the chair nearest each succeeding host's desk, thus sharing the spotlight with show business personalities and with ordinary citizens whose sometimes quirky accomplishments earned them a brief ration of *Tonight Show* fame.

While *The Tonight Show* falls under the Entertainment Division at NBC, *The Today Show*, another Pat Weaver innovation, reminds its viewers each morning that it is "produced by NBC News" from New York's Rockefeller Center. Relaxed host Dave Garroway initiated the program on January 14, 1952, in a 49th Street studio with a large plate-glass window enabling passersby to peer in (and sometimes to be seen on television) and a set decorated with clocks displaying the time zones of major world cities. Searching for ways to snare viewers hitherto accustomed to early morning cartoon and homemaking shows, the producers cunningly added the diapered chimpanzee, J. Fred Muggs, to a cast that included Jack Lescoulie, who handled the sports duties and some interviews, and Jim Fleming, who recapped the news headlines. With the chimp's antics (which included pulling Garroway's already thinning hair and romping around the studio), the program caught the attention of children getting ready to go to school and, frequently through them, their parents preparing to leave for the workplace. Al Raymond remembers that first-cast *Today* era as part of his rushed elementary school days routine in Rhode Island: "It was like, eat your breakfast, get out the door." The program was conceived with the morning rituals of its viewers in mind, with short segments and frequent news and weather updates. In Mary Ann McKenzie's White Plains, New

York, home, her father left to catch the commuter train at 7 A.M., and her mother switched on *Today*, where schoolbound Mary Ann picked up impressions of "Dave Garroway and that monkey" and of Jack Lescoulie: "He had the reddest hair and a bright, shiny face, [for] which on television today they would use makeup to tone down his color. But he had a florid complexion, even though seen in black and white at that time, and you could tell he was shiny on television as opposed to a subdued look."

Although television is often blamed for encouraging the "couch potato" status of many viewers who recline with the remote in one hand and a hearty snack in the other, it has shaped American eating habits in positive ways too. James Beard, revered among gastronomes, offered TV cooking demonstrations as early as 1945, and Dione Lucas followed suit in 1948. Julia Child's *The French Chef*, begun as a local program at Boston's WGBH in 1963, was first seen nationally in the next year and profoundly influenced U.S. food choices and preparation techniques. Penelope Corcoran, restaurant critic for the *Seattle Post-Intelligencer*, recalls her first on-screen glimpse of the woman she would later know "in person": "We were visiting family friends in Short Hills, N.J., one of those leafy New York City suburbs with first-rate public schools and no shops or commerce of any kind. At the very moment we tuned in, Child was, quite literally, whacking a lobster with a cleaver. All the while saying something like 'And so we simply chop! And chop!' In an era of unforgettable TV images, this was one more that an elementary schoolchild was not likely to forget. I certainly didn't."[6] Ann Kirschner notes a very different connection between Julia Child and a viewer: "In my grandmother's East Harlem apartment, noisy with children and the sounds of 110th Street, we knew that when it was time for Julia Child on TV, it was time to be quiet." Although the matriarch's "taste ran more to kreplach than quenelle" and she "could not read English well enough to follow a recipe," her small kitchen produced "an assortment of wonderful food, all kosher and prepared without measuring cups or Cuisinarts." While the grandmother had no intention of duplicating any of Julia Child's recipes, still she "listened to Julia as to no one else. . . . Grandma knew with the instincts of a great cook that she and Julia shared a bond: the pleasure of preparing a dish to be shared with family and friends."[7]

Notes

1. Quoted in Leah K. Glasheen, "Reconsidering the '50s," *AARP Bulletin* (January 1998): 20.

2. Glasheen, "Reconsidering the '50s," 8.

3. Quoted in James Barron, "On One TV Channel, Impostors Get Respect," *New York Times* (January 16, 2002): sec. B.

4. "The Games People Played in a Simpler Time," *New York Times* (October 28, 2001): Arts section, 11.

5. Roger Ebert, *Chicago Sun-Times* (September 16, 1994): 23; www.suntimes.com/ebert_reviews/1994/09/940795.html, accessed July 1, 2003.

6. "Encounter with *The French Chef* Made a Lasting Impression," *Seattle Post-Intelligencer* (August 14, 2004), http://seattlepi.nwsource.com/food/186266_juliaconnection14.html, accessed August 24, 2004.

7. Letter to the Editor, *New York Times* (August 17, 2004): sec. A.

$$—\ 8\ —$$

"Our Town" TV

When newly appointed Federal Communications Commission chairman Newton N. Minow made his "vast wasteland" speech before the National Association of Broadcasters meeting in Washington on May 9, 1961, he did not exempt local broadcasters from blame in the "squandering of our airwaves." Rather, he found, "...too many local stations have foregone any efforts at local programing [sic.], with little use of live talent and local service. Too many local stations operate with one hand on the network switch and the other on a projector loaded with old movies."[1] In some cities, "local" programming consisted of formula homemaking or children's "peanut gallery" shows, inane editorial comment at the end of the local evening news (often repeated at 11:00), or a familiar personality's offering lively brief transitions between segments produced elsewhere and dumped into a time-filling half hour show under a generic title like *Panorama*. Still, some channels developed identities beyond the design of the station break I.D. graphics card, and some local programs, often on exceedingly narrow budgets, bridged well between the individual viewer and network or syndicated offerings. Not all was lost in those hours when no network feed was available. "Back in Philadelphia, where I was born," Marion F. Dilger says, "we had the big New Year's Mummers Day Parade. I can remember my grandparents coming up from Wildwood, New Jersey, just to be with us all day watching it" on a local TV channel.

Serving for more than thirty years as an engineer for WBTV in Charlotte, Frank F. Bateman knew local program successes and surprises as well as many technical challenges before his retirement as Vice President, Engineering, in 1980. His recollections are particularly useful in showing the gaffes that viewers sometimes saw:

Back in the early '50s, all commercials were either on 16 mm film, or on slides with voiceover, or were done live on camera. A live "spot" or commercial was scheduled where the talent, Bob Raiford as I recall, was to pitch some product on a kitchen set. The script called for him to reach up, open a kitchen cabinet, take out the product, and begin his live pitch. When he opened the cabinet door, however, things began falling out in a rush, as if it were Fibber McGee's closet. That's a rough thing to recover from in a live commercial. That is one reason that 99% of commercials are on videotape today.

Another time when the very popular Betty Feezor homemaking show decided to do part of the program live outside from the grassed area behind the new studio building (1955) on Morehead Street, Betty had her young son driving around on the grass in a go-cart. As he circled around again in full view of the camera, he didn't make a sharp enough turn, and he ran right into his mother, knocking her to the ground! Unhurt, she got up and continued with the program. The show must go on, you know.

Life at a broadcast station was a lot of hard work, a lot of fun, and always presented new challenges relating to personnel, operations, equipment, and management.

Complicating the lives of on-camera personnel were "all kinds of bloopers and pranks pulled on everyone," says Garnette Bane, who worked for WTAR Radio-TV in Norfolk, Virginia. She adds, "You can't imagine what the other side of the camera/mike was like!" For the home viewer, the anchor's reading from a teleprompter would seem to be an easy way of making a living, but attention could wander when someone across the studio was dropping his pants as a joke.

Charleston, South Carolina, is not a huge city, but being very distinctive in its observations of history, cuisine, manners, and mores, it is a veritable honeycomb of tradition. Being "fashionably late" to invited occasions is a part of local custom, and whether or not one lives "south of Broad" is an index to family, economic, and social status. It follows that when local channels sprang up in Charleston, they would mirror the city's distinctiveness, and a variety of two- and four-legged beings delighted local viewers of channels 2, 5, and later 4.

Hardest to miss in Charleston's TV menagerie was Suzie-Q from Channel 2, an elephant that lived, according to former news director Ed Webb, "in a garage-sized shed adjacent to the [studio's] back doors and was constantly attended by a thick swarm of flies." On one occasion when the state's governor was expected to hold a live press conference at WUSN-TV, "the studio air conditioning system had taken its annual summer vacation," and before the broadcast, the staff opened the doors "in hopes the night air would cool the place a little." However, Webb continues, "We were no sooner on the air with the governor and a gaggle of state press than the flies descended on the whole group like the locusts on the Egyptians." Everyone in the studio began energetically fanning, and a wide-angle shot showed the reporters "rising individually or in small groups, waving their arms in little wind-milling circles about their heads. It lasted the entire 30 minutes and was a totally fascinating exhibition." Meanwhile the governor "was declaiming on the complexities of the state budget or something equally weighty when a fly flew into his mouth and was immediately swallowed. A stricken look swept over his face, and he grabbed frantically at a glass of water and drank it, appalled that he had been so publicly assaulted. . . . Afterward, he confided that it had been the most depressing single engagement in his political life."[2]

Viewers saw the elephant "live" in parades and in goodwill tours around the state, and she made the newspapers after one escapade or another. Like her attendant flies, Suzie-Q was herself attracted to bright lights, and Charlie Ogletree remembers an episode when she "managed to get loose one night and crossed the highway to a small, brightly lighted gas station and wreaked havoc with the lights."[3] In about 1958 a local columnist complained that the halftime shows at the football games at The Citadel, the state's military college, had become too predictable and needed a circus-like excitement. Members of the Regimental Band quietly arranged with Channel 2 for the "abduction" of Suzie-Q, whose several days' disappearance was the subject of hyped mystery and alarmed speculation. At halftime during the next game, Kathy Hoy Wise remembers, "the gates were opened and here came Suzie-Q being ridden by a bunch of cadets . . . ,"[4] and band member Charles Mitchum adds that the elephant bore a sign that read "something like, 'You wanted a circus; well, here's the elephant!'"[5] Worth Waring comments, "As proof that TV rots one's brain, I can attest that I remember next to nothing of high school math, but I have no problem at all remembering Suzie-Q from Channel 2 . . . a remarkably gentle beast of an elephant that didn't seem to mind being surrounded by a horde of screaming children, at least as long as we didn't run out of peanuts."[6] Rainey Evans, who performed on early children's shows in Charleston, also recalls that Drayton Hastle, the original owner of Channel 2, regularly brought his pack of hunting dogs with him to his studio office.[7]

Meanwhile at Charleston's WCSC, Channel 5, weather man and practical joker Charlie Hall used Roscoe, a pet monkey, on his 3:30 Monday through Friday show *Uncle Charlie's Playhouse* from 1965 to 1968. According to Melvyn Boyd Smith, who as a high school student also worked there, Roscoe "bit everyone in a three-block area centered at Charlotte and East Bay, the Channel 5 studios. . . . He often escaped and we tried not to find him." Melvyn Boyd Smith also remembers that Charlie Hall had "all the ladies from Channel 5" eagerly expecting a studio appearance by singer-actor Pat Boone, who was an apparent no-show until someone glanced at a calendar and noticed that the date was April 1![8]

Soon after Channel 5 began broadcasting on June 19, 1953, Annie Lee Smalls hosted a Saturday morning program in which she "dressed up in a lovely blue dress and read stories to children," according to Rainey Evans, who adds, "Since we were 'live' in those days and Annie Lee had to have a Saturday off every now and then, Charlie [Hall] would fill in. Once he donned the 'lovely blue dress,' put a cigar in his mouth and read as though nothing was amiss. Needless to say, Charlie was in pretty hot water come Monday morning."[9]

Another distinctive sight for Charleston viewers was the "infamous incident" on Channel 2's late local news, as watcher Fred L. Smith recollects it: "At the very end of the live broadcast . . . the chubby newscaster stood up to reveal he was in his underwear. That's right—no pants, just polka-dot boxers! He stood there, motionless, for at least 15–20 seconds before the cameras were shut off."[10]

Melvyn Boyd Smith was in the studio that night and corrects the widespread misimpressions of why "the handsome and talented Jack McKee" was in his underwear behind the anchor desk. It was not true that this was a necessary response to hot lights and failed air conditioning on the set; rather, "It was that Jack had to do a live Builders (a Lowe's kind of store) commercial just a moment or so after the newscast that required him to don a stocking cap and nightgown and, holding up a candle, relate the benefits of [the] Builders 'Midnight Sale.' To facilitate a quick change he had already doffed his pants" at the director's suggestion. "Live commercials were always an adventure and what transpired when Jack caught a look at himself [in the monitor] sans pants was memorable." He turned "red as a beet," tossed the commercial copy in the cameraman's face and, "carrying his pants, he left the building!" Contrary to a wide public impression, the announcer was not fired—the director was blamed for pressuring him on the quick change of apparel—but he soon resurfaced on another channel in another town.[11]

Teaching communications courses at the University of Miami when television came to South Florida, Corinne Holt Sawyer quickly learned first-hand the economies, exigencies, surprises, and general risks of local programming:

Because Miami had no co-axial cable for a long time (four or five years at least) after they started broadcasting TV, and because they obtained network shows on kinescope, which was of poor quality, and because the entire day's programming from the networks was not available locally, Miami stations were anxious to find ways to fill some of their daytime hours. WTVJ, with studios in downtown Miami, negotiated with the University of Miami's Department of Radio and Television to do a regular half-hour show throughout the school year.

I'd had training in both stage and radio directing, so I ended up in charge of the majority of the university's first year of TV broadcasts. Incidentally, my husband and I had no television of our own, and we didn't even know anybody who owned a TV set, so we had to go to a local bar to watch the productions staged by the other directors—but we did it, partly because having nobody to teach us TV techniques, and having no TV to watch the professionals at work, we had to learn by doing and by watching each other.

We presented several half-hour dramas, and the hazards of performing live on TV (especially with student actors, inadequately rehearsed) made for occasional disasters I remember one show in which, about five minutes before the close, the stage-right wall of the set fell over on the actors, and the cameraman, with great presence of mind, simply panned away from them. Puzzled viewers could hear the dialogue continuing without a break (we'd drilled the actors in "the show must go on") but were shown only a couch, a chair, an end table, and a vase—no live actors. They were at the other end of the set reciting their lines while they tried frantically to prop up the rest of the walls, so the last bit of the drama could be visible to the audience, as well as just audible.

We alternated the playlets with a program featuring a faculty member from the university, and often from one of the science departments, because of the potential for demonstrations, as opposed to a straight lecture. A political science teacher charted the demographics of Miami (this was years before the "Cuban invasion.") A meteorologist showed us how hurricanes form (using a lot of charts, as I recall). A psychologist showed live white rats learning to run a maze. A home economics teacher showed us how best to stuff a turkey. A zoologist brought live snakes, one of which escaped in the studio. (The camera work became suddenly very wobbly.)

And then there was the termite expert. He brought with him a desk he wanted to use as a table to hold his props during the show, which we considered eccentric but harmless—and he showed us live termites under glass, which we were able to magnify to fill the screen with highly satisfying visuals; he showed us a piece of termite-damaged wood, so we could view termite tunnels in cross section. And as the program wound to its close, with no warning the soft-spoken expert grabbed up an axe we didn't know he had, and he attacked the desk: *BLAM—BLAM—BLAM.* He chopped that desk into kindling, and I concluded that he had lost his mind. I was about to tell the camera crew to run for it and fill the airtime with slides and music. Then, just as suddenly, he stopped hacking and held up one of the broken legs to show us that the desk was riddled with termite tunnels. He wanted to demonstrate how hopeless it was to salvage such a piece of furniture after the damage was done. We were able to finish out the show in the regular way, but my heart was racing.

Several local programs held Marilyn Chadwick's attention in Sioux City, and she appeared on TV herself more than once, in person and in imagination:

We had one station, KVTV, Channel 9. There weren't many local features, apart from news, weather, and sports. We welcomed these commentators into our home like family.

I'll never forget Charlie. As we, and the community at large, watched him deliver superb newscasts nightly, we also couldn't help noticing a phenomenon unfolding rather swiftly before our eyes. His hairline was receding at record pace. Then one night he appeared with a full head of hair swooping across his forehead. Nothing was said. The viewers were not prepared. However, the next day reputable barbers and hairdressers displayed signs in their windows which read, "HE DIDN'T GET IT HERE!"

Another employee of Channel 9 was Jim Henry. One of the many hats he wore in the '50s was a cowboy hat; he was the Canyon Kid on a show called *Kids' Corner.* This show was "live" after school at 4:00. The set included bleachers for visitors and guests, all of whom were interviewed by Canyon. My sister and I were thrilled to be guests on the show twice. The first time was Pet Day, and we brought Rusty, our cocker spaniel, who was miserable under all those lights and didn't care for the close proximity of all the other animals present.

Canyon's special guest that day was The Range Rider. He starred in one of my favorite Westerns of that era, so this was my first encounter with a genuine big star. He gave each of us his autograph, and I stood close enough to him to get a whiff of the strong, rich aroma of his leather cowboy outfit and boots! VCRs being a long way off in the future, my sister and I had to rely on our mother's biased report as to how well we looked on TV.

Another local feature was a show presented by a lovely lady by the name of Jan Voss. Her set was a kitchen table, behind which was a sink and cupboards. She would whip out recipes each day and explain, step by step, how to turn out these culinary delights. She was my idol. I cast aside my aspirations to be an elevator operator when I grew up and decided instead that I wanted to be a homemaker. Our kitchen became my set, and I would pretend I was on TV. I made everything from pudding to poached eggs and became rather adept at cooking while facing one direction only, which was necessitated by my need to always face the camera, that being the kitchen window, which framed me appropriately.

There was another local program which was a game show. It was called "VIDEO." The cards were picked up at participating grocery stores, and the game was similar to bingo. We had enough cards to practically cover the living room floor. Mom was a whiz at winning. When a row was filled on a card, she had to call the five-digit telephone number to win. Other would-be winners were also dialing, so the lines would get tied up with a busy signal. She would dial four of the digits ahead and the fifth when she completed the row, in order for the call to go through. One neighbor was irate with my mom and said she was cheating. Nevertheless, Mom continued to rake in those prizes.

We had a local weatherman who really stood out among the rest. Since his entire report consisted of pointing out different things on the map behind him, he made the weather interesting by using a different pointing device every night. Viewers participated by sending in different things, and their contribution was always acknowledged and their names mentioned on the air. An umbrella might be used for a rainy forecast, a candy cane on Christmas, a golf club, a pool cue, a walking stick, a sword.... It transformed a mundane weather report into FUN!

Marilyn Chadwick ends her recollections of local programming with an impression of the early studio. "The props and sets weren't nearly as attractive as they appeared on camera," she says. "Also, there was nothing cozy about Jan Voss's kitchen. It was a little corner of a very big room that had cement floors and was very drafty, kind of like a big warehouse. It was in one of the oldest buildings in the city, and everything was primitive and reflecting of a low budget. Quite a contrast to the costly studios of present times."

One low-cost form of local programming developed in the 1950s was the dance party show on weekday afternoons or Saturday mornings. A local Philadelphia show, *Bandstand*, began in 1952 with radio disc jockey Bob Horn as host, and he was succeeded by Dick Clark (also from radio), who five years later took the program to ABC-TV under the title *American Bandstand*. The network show encouraged a new round of local efforts, such as *Dave Senate's Dance Party* in Providence, Rhode Island. Future engineer Al Raymond and a girl from his high school were selected to appear about 1963, before he had a driver's license: "My dad had to drive us down to the studio in downtown Providence; it took maybe an hour and a half or so. It was filmed at about 10:00 on Saturday morning and televised later, in the afternoon, so we got the chance to see ourselves on TV. They did interview people that were there—they asked who we were and what school we were from—but it wasn't a competition as such. It was just ... they played music and everybody danced in couples, you know, fast number, slow numbers.... At high school reunions we still talk about those shows." He was "really surprised" by the appearance of the studio: "When we got there, it seemed so small compared to the way it looked like on TV. With all the light in the overhead and stuff . . . you don't see that stuff on TV."

An unpleasant surprise for one-time or occasional guests on local shows was the makeup required. Owens Pomeroy remembers his Baltimore experience: "Makeup in those early years was atrocious: gray base powder, black nail polish, and brown rouge and lipstick. If you saw these people off camera in person, you would think they were made up for a horror movie." He notes that when former radio personalities such as Milton Berle, Jack Benny, Fred Allen, and George Burns brought their programs to television, they were "appalled at the grotesque makeup they had to endure," and they made jokes about it on their shows. Guests "were shocked, even to this day, how a TV picture makes you look about 10 pounds heavier. I remember being on a local variety show, where I did a lip-sync act, and when I saw the kinescope, I was flabbergasted. I looked about 10 to 20 pounds heavier, and being a big man of over 220 at the time, my friends told me I looked like a butterball on TV. [Even with] all the technology they have today, this distortion problem still has not been completely solved." Beyond the indignities of makeup and the camera's effect on body appearance, the early years brought wardrobe restrictions. Guests were warned not to wear white blouses or shirts, which produced a strong glare in the black and white cameras. Luckily, light pastel pinks, blues, and yellows were fashionably "in," and in the TV picture they registered satisfactorily as white or slightly off-white.

Although he claims no "on camera" appearances of his own, Harold Woodell fondly remembers two local "low-budget affairs that came on in the afternoon," sandwiched between "the usual game shows, old movies, and variety shows" on Channel 2 from Greensboro, North Carolina:

> One of them was a show featuring a folk singer whose name I still remember, Forrest Covington, who played a guitar and sang old ballads like "Greensleeves" while sitting on a stool. He dressed in flannel shirts buttoned at the neck and did his "act," if you can call it that, in plain vanilla fashion. I have loved folk music ever since.
>
> The second show was one that played popular music of the day while an artist drew an illustration on a sheet of paper tacked to an easel. I never saw the artist, only his or her hand that was creating the drawing as I listened to the music.
>
> Thus, if the song was "Shrimp Boats Are A-Coming" or "You Belong to Me" with the opening line "See the pyramids along the Nile," the artist might draw a scene in which shrimp boats would be pulling into port or the Nile River with pyramids in the background. I was the only member of the family who thought this act was fascinating.

The artist's sketching show also brought Harold Woodell a moment of disillusionment. Having spent a good portion of his time in the public library, he had looked many times at Bill Mauldin's book of World War II cartoons, *Up Front*. "I knew many of those pictures very well, almost pen stroke by pen stroke," he says, and one day a guest on the art program displayed copies of those cartoons, claiming them to be his own work. "It was a sad little bit of fraud," the viewer observes.

During the 1950s, Malcolm Usrey rented a room in a couple's home in Post, Texas, and, after attending to lesson plans and other aspects of his public school teaching duties, he often joined his landlords to watch television. *The Bell Telephone Hour* and Spring Byington in *December Bride* were among his network favorites, but his most memorable viewing moment came one Saturday morning when a local host on a Lubbock channel was interviewing a preadolescent boy in Art Linkletter "Kids Say the Darndest Things" style. The first question established that the boy's father had been away for some time on military active duty. "Well, then," the host asked, "do you still sleep in the bed with your mother to protect her?" "Yes, sir," came the response, "except on Saturday nights, when my uncle comes over to do it." The station's picture went instantly black, and the audio was abruptly silent. "And it stayed that way for quite some time," Malcolm Usrey recalls. In those censorious early TV days, the blank or black screen was the broadcaster's electronic equivalent of blushing. Bob Hope's jokes and Groucho Marx's offhand double entendres occasionally prompted "Program interrupted" cards to pop up during their network programs, but these were expected by audiences and network officials; they were part of the game, aces up the sleeve to be played at just the right moment. Local programs, taking without-a-net risks with unseasoned guests, were far more vulnerable to awkward moments and blank screen "time-outs."

Did Mr. Minow say "vast wasteland"? Well, among local programs there were oaises, too—and mirages—for the beholder's eye.

Notes

1. "Excerpts from Speech by Minow," *New York Times* (May 10, 1961): sec. A.

2. Quoted in David Farrow, "Story Involves a Governor, an Elephant, the Press and Flies," *Post and Courier* (Charleston, SC) (July 3, 2003): sec. D.

3. Quoted in David Farrow, "Reader Recalls Short Story Has Suzie-Q Sent to Atlanta Zoo," *Post and Courier* (October 3, 2002): sec. F.

4. Quoted in David Farrow, "Happy Raine Shares Her Memories of Early Days of TV in Charleston," *Post and Courier* (October 10, 2002): sec. A.

5. Quoted in David Farrow, "Elephant Antic Was Set Up, Reader Says," *Post and Courier* (October 24, 2002): sec. D.

6. Quoted in David Farrow, "Remember When? Children Delighted in Suzie-Q," *Post and Courier* (August 29, 2002): sec. F.

7. David Farrow, "Happy Raine Shares Her Memories of Early Days of TV in Charleston."

8. Quoted in David Farrow, "Reader Recalls Early Days at Channel 5 with Charlie Hall," *Post and Courier* (November 7, 2002): sec. F.

9. Quoted in David Farrow, "Happy Raine Shares Her Memories of Early Days of TV in Charleston."

10. Quoted in David Farrow, "Local Newsman Airs Polka Dots on Broadcast," *Post and Courier* (November 14, 2002): sec. F.

11. Quoted in David Farrow, "Truth Told About Newscaster in Boxers," *Post and Courier* (November 21, 2002): sec. F.

— 9 —

The Distant View

"Often the way to know the personal value of something is to do without it for a time." "Yearning for something often leads to fall-on-one's-face disappointment when the apparent desire seems finally fulfilled." Even if those statements fall under or near the category of truisms, they are validated by the experience of American citizens living in other countries during the early television era and by the forced reassessments of immigrants to the United States.

Warren Edminster spent two segments of his early years living in Africa with his Baptist missionary parents, with an interim year back in the United States in 1977:

I was in Rhodesia (now Zimbabwe) from 1969 to 1977. We were so distant (supplying our own electricity, water, bricks for buildings, and so on) that we had no television. Traveling to the cities was a magical experience in part because we looked forward to watching television, and we even planned our trips around certain weekly shows. I remember on one trip to Salisbury (now Harare) that we anticipated with delight being able to watch *Desert Rats*, a show we had enjoyed before about British Raiders in North Africa during WWII. When we got to the missionary guesthouse in Salisbury, we sat around the telly just waiting for the show to come on. The time came and went, with no *Desert Rats*. As our dismay mounted, Dad called the TV station and was told that *Desert Rats* was no longer on the air. The disappointment absolutely ruined the entire trip.

We lived in Rustenburg, South Africa, a small mining city of 22,000, from 1978 to 1982. The government broadcast TV for only three hours a day, 1.5 hours in English and 1.5 hours in Afrikaans. It was a surreal experience to watch J. R. on *Dallas* speaking in gutteral Afrikkans.

Culture shock was a factor both times [on returning to the U.S.], but it was more pronounced in '77, when I was 11. I became a bit of a TV junkie for that year.

Warren Edminster recalls that on reentering the United States, "I was overwhelmed by the continual rush of sights and sounds." After another moment's reflection he adds, "The media made the difference," and at the center of that sense of change was television.

Alina Bales, now living in a Lutheran retirement center in Carlsbad, California, came to know U.S. television by a very different route. Born on a farm near Milan and living there and in the city during World War II, she never even heard of television during those hard years. Accompanying her sister and her ex-fascist brother-in-law to Caracas, Venezuela, with the hope of a new start in life after the war, she found her existence there almost as difficult as it had been in wartime Italy, for the Venezuelan government was in turmoil, and there was revolution—shooting in the streets, threats, buildings burned, public demonstrations. She did, however, marry Robert Bales, an American working at the embassy in Caracas.

Venezuela had no television at that time, and neither, for that matter, did her husband's family in Ardmore, Oklahoma, when the newlyweds visited them there in the 1950s. In 1960, though, Alina and Robert Bales settled in Washington, DC, and she looked forward to one of the things she had understood that every U.S. home had, a TV set. "I expected it to be my window on my new world," she says, but there too she met disappointment:

> This new country where I was going to live was going to furnish me the peaceful life I'd never known in Italy and in Venezuela. And I was so happy and excited to have that TV.
>
> But, did you know, first there was the Kennedy assassination, then the Vietnam war and the Johnson era with all the protests . . . you know, Jane Fonda, the student organizations like the Weathermen, and the shootings at Kent State, the Symbionese Liberation Army. . . . I don't remember them all. But my wonderful TV, my window on my new world, showed me nothing but anger and destruction, people being shot, people shouting curses, people marching in protest.
>
> I was so disillusioned! I wished we'd never got a TV at all! I was naive, I know, but I blamed TV for all the upset of the world around me, for all my disillusionment in my dream of a perfect, peaceful society. And even now, all these years later, I can hardly bear to watch it.

Mystery writer Corinne Holt Sawyer, who generously transcribed Alina Bales's story of an unhappy acquaintance with television, had a happier experience of her own when she and her husband were living in England in 1955. She saw some programs that would later delight American public television viewers, and she discovered some striking differences in the pace and content of British television:

> It was common there to rent, rather than to own a set, and we rented a 15-inch black-and-white cabinet model. Independent Television didn't begin broadcasting till the last year we lived in England, but we didn't mind—the BBC had three networks. They often began broadcasting late in the day, rather than all day long, and "starting time" varied depending on what they had to show. To fill the time they did have, they

gave us an awful lot of news shows, both hard news and features, and some wonderful documentaries on nature, on history—wonderful stuff, plus unexpectedly interesting sports programs like the horse-jumping trials. It was like being able to watch an expanded and comprehensive PBS on all three BBC networks (while Americans were watching Uncle Milty and *Your Show of Shows*—and from what I've seen of them on reruns, I didn't miss much).

One of the BBC networks had an excellent "magazine show" devoted to features and to news in depth; it aired just before the evening newscast. My favorite, and the one that I recall the best, was a feature they did about the spaghetti harvest in Italy. They showed peasants picking the nearly ripe strings off small, gnarled trees, from which the spaghetti hung thickly on every limb. The documentary went on to tell how they tested for ripeness, how they stored the newly gathered spaghetti while it dried, so that it could later be packaged.... I was completely mystified. The BBC goes to extraordinary lengths to authenticate the content of their shows, so I suspected that perhaps there really was a kind of spaghetti that I didn't know about that was grown instead of made, and I didn't catch on until the program closed out with its usual final announcement—in which they always mentioned the date—and on this occasion pointed out (without special emphasis) that today was April 1st.

The British are stereotyped as having no sense of humor. But in fact, their humor is rich, though generally more subtle than American humor—often both dry and sly—and the BBC is no exception. An American network wouldn't have done that spaghetti feature—but if they had, they'd have messed it up with explanations of and apologies for the hoax—before, during, and after.

One of my favorite things about the BBC in the 1950s was that they didn't fuss with exact timing. I was trained to work in American broadcasting, where we made our programs come out to the exact second. But things were very different on the BBC. If a program ran over by a few seconds or even, in a few cases, a couple of minutes, they simply notified the other control room to delay the start of the next program. And if a program ran short, whether by seconds or minutes, they had a few stock films that they ran to fill the time. They had tropical fish in an aquarium; they had the waves lapping on the ocean shore; they had a clear, babbling brook running over rocks and between grassy banks; and my two favorites—in one, they had a potter working on a wheel, drawing up a tall vase, squashing it down and starting over to draw up a bowl; and in the other, they had photographed two kittens playing energetically with some balls of yarn.

Sometimes the fillers made better watching than the programs, and I surely wish some of the American networks would use the same strategy. There are lots of programs that could stand to be cut in length. But knowing American TV, they'd probably fill the time with commercials for mobile homes, accident lawyers, and used car lots. Of course in some cases that would be no bad thing. I've seen a lot of commercials that were better done and more inventive than the programs on which they appeared.

Marilyn Chadwick, living in West Germany in 1964 and 1965, found that being away from U.S. television quickly led to gaps of understanding as well as bouts of nostalgic longing for at-home viewing habits. "While in Germany," she says, "I was very homesick, and what I missed most was sitting down as a family every Saturday night for *The Jackie Gleason Show*." Her family rented an upstairs apartment in a small German village, and the landlord headed one of the very few families in the vicinity to have a television set. She recalls, "They had *one* program *only* from the States, and they loved it: *Bonanza*. The English dialogue was muted, and the German language was dubbed in. We found it to be hilarious. We also learned from that experience that *Bonanza* was responsible for forming an impression of Americans that we all dressed in boots and ten-gallon hats and rode horses across the Plains. We were a nation of cowboys! We set our German friends straight but could certainly see how they arrived at those conclusions."

One of her mother's letters from the United States brought Marilyn Chadwick a new puzzlement:

She asked me if I had heard about the Beatles that had come over from England. I didn't have a clue what she was referring to but wondered if there was an invasion by these insects as devastating as was the grasshopper plague spoken of by my great grandmother. In a subsequent letter Mom told me that the Beatles were going to be on Ed Sullivan's show. Now that my theory was shot, I had to do some thorough investigating to find out who or what she was talking about. Not being exposed to a television in our home those two years, I didn't get my first eyeful or earful of the Beatles until 1966.

$$—\ 10\ —$$

The TV Life

In some homes, the TV set was an alien thing, turned on and off at specific times for specific programs, but most of all kept in its place. In other settings, it played all day long, gathering the upward gaze of moppets on the floor and hawking aspirin, tires, and *The Days of Our Lives* to anyone passing by. For military families and others who moved often, it provided a sense of continuity, and for young viewers it was often a refuge from unhappy school experiences or parental bickering. It suggested career paths to some, shaped life-views for others, provoked moral or aesthetic outrage, or encouraged a firm hold on truisms. Pastors found sermon topics in *Life Can Be Beautiful*, and pilgrims still flock to shrines celebrating *The Andy Griffith Show*, *The Waltons*, and *I Love Lucy*.

Few ties to television can be more intimate than carrying through life a name taken from a TV character or another facet of the medium. Mrs. D. T. McDonald, at the urging of her mother and her sister, named her daughter Darla for a character in the *Our Gang* comedies that became a staple of early television. Louise F. Glick took a cue from the corporate name of Lucille Ball and Desi Arnaz's TV production company:

In the fall of 1954 we were expecting our third child. In those days many of our favorite programs were Desilu Productions. My husband, Ivan, and I (Louise) thought it would be fun to send our birth announcements [describing the baby] as the latest "Ivalu Production." Our baby arrived early, and the announcement was still only in our thoughts. When she was five days old—in those days mom and babe stayed in the hospital five days—the personnel wanted a name for her birth certificate. This was our third daughter, and we couldn't agree. So we decided to name her after us as Iva Louise and call her Iva Lu. I'm not sure that she likes her name, but our intentions and memories are fond ones. She's a very special daughter!

"I was born in 1953 and raised in Brooklyn, New York, as an only child. Hence, the TV was my very best friend," says Tom Worsdale, who adds, "I bet I could have won just about any trivia contest that dealt with television between 1960 and 1970, and I have a few memories from the '50s too." He saw the premiere episodes of classic sitcoms and "became a regular watcher of the ones I really liked [including] *The Beverly Hillbillies, Green Acres, Petticoat Junction, Andy Griffith*, and *Gomer Pyle*—living in NYC, I liked all this country cornpone—*The Flintstones* (a Friday night show when it started!), *Bewitched, Gilligan's Island, The Time Tunnel, Land of the Giants, Outer Limits, Wild, Wild West*. All these shows really stood out in my memory, and being an only child, it was just me and my best friend, our little twelve-inch black and white in a big box."

Now in his mid-fifties, a high school art teacher declares, "My brain has been a storehouse of seemingly worthless information until now. I can recall almost anything involving TV of the 1950s and '60s. . . . I grew up with and lived it! For example, I know that Vern Albright (Margie's father in *My Little Margie*) worked at Honeywell and Todd, William Bendix in *The Life of Riley* worked at Cunningham Aircraft and lived at 1313 Blue View Terrace, and the Kingfish's name on *Amos 'n' Andy* was George Stevens, and he was a member of The Mystic Knights of the Sea lodge."

Luanne Hege says, "TV viewing was very important to me as a girl. I started avid TV watching in the late '40s, became a real devotee in the '50s. With Daddy gone so much and our hop-scotching around the nation, it was a 'friend' to me— steadfast and constant." Rather than centering on the details of programs watched, her memory focuses on early receivers: "Our first TV was a Muntz, hawked on the tube by 'Mad Man' Muntz. It had a funny console with the lid that you raised and the TV picture would be reflected in a mirror. I also remember having an Olympia TV. The reception problems were enormous. Depending on the great TV gods, I might sit down to watch a show and be treated to the rolling picture, with various zigzag patterns intermittent."

"When I was a kid," Patience Sibeal Wieland recalls, "most of my fantasy life and play revolved around the things I saw on television or heard on the radio. The weird thing is that the things I most remember—and today often enjoy the most—were things that at the time terrified me. Apparently I used to be afraid of Soupy Sales. I know he was loved by millions of children, but every time he came on the tube, I screamed and screamed until the TV was shut off. I guess I was afraid he was going to throw a pie at me." Other aspects of her early relationship to television were more positive or useful:

As a little girl I was infatuated with half the male population on TV: Captain Kirk, Buck Rogers, BJ of *BJ and the Bear*, Dirk Benedict and the blonde guy on *CHiPs*. (I didn't like Erik Estrada; he had a very insincere smile.) I was not allowed to watch *The Dukes of Hazzard* (for which I now thank my mother), *Sanford and Son*, or *All in the Family*. I did grow up watching a lot of Canadian television: I grew up near the border and thus saw a lot of CBC staples. *Mr. Dressup* (a sort of Canadian

Mr. Rogers who wore glasses, was soft-spoken, had some sort of a trunk with dress-up clothes and puppets) was one of my favorite shows, and I also learned a lot from David Suzuki's *The Nature of Things.*

Our home was not a happy one when I was growing up, so I think television was more than just eye candy to me. It was a form of escape from my parents' crumbling marriage, just as books, music, and playing outdoors were. I watched a lot in the evenings after preschool and later grade school, and today I definitely fit the stereotype of a "media-savvy Generation-Xer."

Parents "back then" often complained about the difficulty of getting their children out into the fresh air for exercise when the youngsters would prefer passive viewing from the couch. "I was definitely a television junkie for the better part of my life," Dixie Ann Schmittou admits. "But I was also an avid reader. To accommodate both of these addictions, I learned to read and watch TV at the same time. I remember all the popular shows, but my favorites were those like *Playhouse 90* and *Armstrong Circle Theatre* plus all the mystery-detective series through the years." In the early days, she adds, "We were so hooked on TV that even the movies didn't compare. Our little town had a very prosperous movie theater, but as TV became more popular, the theater lost a lot of customers and finally closed. TV became our window on the world, and we absorbed as much of it as we could."

When Dixie Ann Schmittou grew up, married, and had children, she found television to be "even more invasive" with its news of assassinations and the Vietnam War:

> The '70s changed television as it did everything else. It was then that people began to pronounce that sex without marriage was more than OK, and now is it any wonder that even pre-teenagers are sexually active when we see every sitcom and almost every drama presenting it as the norm? The families presented make fathers (if they are around) look like total idiots. I'm afraid I ignored a lot of what I should have objected to on TV. But I had boys, and they mostly watched adventure stories. I was the one who watched the soaps and junk TV. It even influenced my reading, and I found myself reading the romance novels. I've given up those too.

Television was responsible for many affectionate and frolicsome moments in Marilyn Chadwick's family life:

> There were very special times I spent with my dad. He and I were the night owls of the family, and Mom and Diane often went to bed early. When I was about nine years old, I would sit in his lap in the big brown chair with a bowl of popcorn and watch wrestling. I didn't understand the rules or anything, but what I did understand was what a bunch of characters those early entertainers were. Dad and I watched and roared. My two favorites were Yukon Eric and Gorgeous George. Gorgeous entered the ring in such a proud and arrogant manner, dressed in a long, exquisite velvet cape and wearing his silver blond hair in soft curls. (Remember, this was the era when men and boys wore butch haircuts and flattops.) He would throw silver hairpins

> from his locks to the women in the audience whose hands were stretched forth in anticipation of being a lucky recipient of these souvenirs. Well, I was a recipient as well! We sent for a hairpin through the mail, and I had so much fun taking it to "show and tell" at school and revealing the source from whence it came!

When the family viewed music programs together, her father could be as entertaining as the on-screen artists:

> There was Liberace on the piano and his brother, George, on the violin. A candelabra sat upon the grand piano, and Liberace always wore elaborate tuxedos and jewels. Dad did a really fine impersonation of him flipping his tails before he took his place on the bench. Dad also impersonated a lady by the name of Carmen Miranda. She used to be in some of the old late movies—a singer and dancer. Her costume usually required quite a headdress—kind of a turban with a bowl of fruit piled on top of it. Dad would toss a couple of bananas on his head, click his castanets, and dance up a storm.

Marilyn Chadwick also notes the influence of television on her sense of fashion and of style. In fact, she says, "Some groundwork was firmly laid for a powerful future influence in this category." She explains, "I learned how to sew at an early age and would often select fabric and patterns in harmony with the styles of my favorite characters. Prior to television, Mom and I relied on the *McCall's* magazine to serve this purpose, but a photograph or illustration was not as effective because you couldn't grasp the dimension; nor could you see for yourself how different styles would drape or swirl. I remember making my evening gowns for homecomings and proms and wondering if I could capture just a shade of the elegance displayed by Loretta Young when she made her grand entrance at the beginning of her show. What a presence! What a lady!" Thus, she says, "This influence in design also affected the way I chose to decorate my room—and later our home. All kinds of ideas were lifted from the living rooms of sitcoms and soap operas."

Even in this area too she sees a shaping family influence beyond the issues of fabrics and seams: "I was always happy to have the guidance, moral or otherwise, of both of my parents. Dad had a subtle way of helping me with my choices. For example, if an opera singer appeared wearing a revealing formal with a plunging neckline, Dad would comment by saying, 'Isn't that an interesting dress she's ALMOST wearing?' I listened; I obeyed; I kept my creations modest! Perhaps I'd be an ideal spokesperson pro parental guidance in children's TV viewing of today!"

Clearly television played a large part in Marilyn Chadwick's family life, but hers was not an uncritical soaking up of everything she saw on the small screen:

> The TV encroached upon time that might otherwise have been spent reading a good book. When I was in the fourth grade (1953), I could hardly wait to get to school in

the morning, because our day would begin by the teacher's reading a chapter in the life of Laura Ingalls Wilder. This was an EXPERIENCE. *I* picked the wildflowers; *I* walked through the tall prairie grass; *I* felt the warm summer breeze caressing my face; *I* was Laura. Later, when *Little House on the Prairie* became a TV series, I looked forward to it and loved watching it, but something was lacking. I had been usurped by Melissa Gilbert as the lead character and was robbed of the privilege of being Laura. This vicarious experience just didn't cut it for me. Things were not the same as integrating myself into the pages of the book. From that point on, when any book was made into a movie or TV series, I found it somewhat of a letdown. I think the innovative imagination provoked by reading is a valuable exercise of the mind, and the mind ought not be deprived of this use.

Teresa Williams, Marilyn Chadwick's daughter, was moved to make two observations when she first read of her mother's life with television (related in the paragraphs above and elsewhere in this study):

> I am particularly interested in her stubborn distinction between what was real and what was make-believe—even to the point of not appreciating cartoons, being INSULTED by them, in fact. The lines seem to have become more and more blurred for children as they are now raised in a TV culture. But my mother, who certainly was not immersed fully in the same TV culture (especially in terms of hours viewed per day), recalls culture more firmly grounded in reality and because of this rejects much of what she sees on her TV.
>
> The other notable idea that she seems to repeat is that TV taught her what her role should be as a woman. Here she is not as eager to reject what she [saw] on television, perhaps because it glorified the [homemaker] role. (If she could only get that pot roast recipe just right, home would equal harmony). I definitely saw my mother in a new light after reading her observations. She was ranked in the top five academically when she graduated from a very challenging and very large high school in Sioux City. She was the "most likely to succeed" person in the yearbook, yet she decided that the career for her was to be a homemaker, and she never attended college. In some way, I now understand that choice.

Shelby Duncan, whose work title is Administrative Assistant, took her career cue from Della Street, Perry Mason's legal secretary, although she learned later that such a position did not pay as well as anticipated. University nursing school instructor Linda Howe, PhD, RN, CS, points out that she was "named after a character on radio—Pepper's wife on *Pepper Young's Family* was named Linda— and she was a nurse . . . so here I am, Linda, the nurse." Having that tie, she has observed television's portrayals of nurses and of the medical profession in general over the years. "I still miss *M*A*S*H*," she says; it "was medically fairly accurate," but it "did damage nursing" by portraying "stereotypical sexy nurse things" in the pinchable operating room assistants and in Loretta Swit's role as Major Margaret ("Hot Lips") Houlihan, the nurses' commanding officer. Among later series, *China Beach*, set in the Vietnam War, "was good and should have stayed on" despite some stereotyping of nurses, "and *LA Doctors*—it had promise. Too bad the writers didn't. The horrible one was *Nightingales*," Aaron Spelling's medical series, which lasted four months in 1989 before succumbing to general critical disdain and strong protests from the nursing profession. "Nurses worked hard to get that one cancelled, and we DID," Linda Howe comments. Of *St. Elsewhere*, set in the Boston hospital more correctly called St. Eligius, she says, "The medicine was fine, nursing was weak. *General Hospital* is and was ridiculous, as was *The Guiding Light*, although I didn't know that at the time at Grandma's house!"

Juana Green is a scholar and recently a university teacher of Renaissance English drama, with special interests in material history. Her life would have taken a different path, however, if she had fully followed the career path that her mother envisioned for her:

> When I was growing up, my mother's favorite show was *Lawrence Welk*. And as a result, or at least this is what I believe looking back, much of my education was constructed around *The Lawrence Welk Show*. So I had the privilege of taking accordion lessons so I could be like Myron Floren, Welk's accordion player. And I also got to take ballroom dance lessons so that I could be like Bobby and Sissy King. In fact, she had me take lessons at the same studio where Bobby and Sissy King took their dance lessons in Long Beach, California. However, none of it worked out in that I was never too enamored of the accordion. Toward the end of grade school I got really involved in [watching] television shows, and when they changed the day of *Lost in Space*, I tried to convince my mother that I should drop the accordion lessons because I really wanted to watch this television program at that time. However, she would have *none* of that because (laughter here) I needed to go to the accordion lessons so I might become a famous accordionist someplace, I'm not sure, playing weddings or something.

In social terms, the ballroom dancing lessons were more profitable, but "Once I had to start going to ballroom dance contests, that was pretty much the end of it, because it became clear how klutzy I really was as a ballroom dancer. So my mother clearly had inflated visions of my dancing abilities, but still, most mothers do, right? That's what I remember about *The Lawrence Welk Show*."

Tom Maertens, Sr., according to the son who inherited his name, "was a graduate of West Point and a career officer." For that reason, says Tom Maertens, Jr., "...my brother and I were granted exemptions to the 'go to bed by 8:30' routine in order that we watch *West Point* [a 1957 anthology series about life at the Academy] and *The Gray Ghost* with Tod Andrews [another 1957 series, this one set during the Civil War]. Interestingly, I graduated from West Point, and my brother and I had careers in the Army. Go figure. Other shows that influenced me were shows on fire, police, and rescue, such as *Rescue 8, Emergency, Highway Patrol* with Broderick Crawford ('21-50, bye'), *Dragnet, Tales of the Texas Rangers, Sheriff of Cochise, Sky King* ('Take her down, Penny,' aviator Sky says to his niece at the plane's controls), and *Whirlybirds*. I ended up flying helicopters and airplanes in the Army and have spent over 20 years as a volunteer firefighter and EMT."

Advertising freelancer Ernie Schenck recently examined his attitudes toward his profession and noted that he had never felt the desire to join the choruses of awards show judges who will often praise an ad campaign's imagery, slogans, or concept by saying "I wish I had done that." Schenck notes that he has *admired* many campaigns and has seen some that "literally took my breath away," but "I just never wished I had DONE it." Then, scanning an old issue of *Psychology Today* while he waited at the dentist's office, he came across an article which reminded him "that many psychological issues can be traced back to a single event in our lives," often at a very young age. That thought led to an epiphany as he recalled his interaction with *Winky Dink and You:*

Winky Dink was this weird little character that had a head shaped like a star and went around getting chased by lions and crushed by falling boulders and stuff like that.... OK, so the premise wasn't exactly mind-popping. But here's the cool part. Every week, Winky Dink would have an accomplice. A sidekick. Me!

No, I wasn't on the show. But I *was* on TV. Literally. See, what you did was, you sent away for your official Winky Dink Kit which consisted of a special transparent vinyl magic sheet, some special magic crayons and, of course, a special magic eraser.

Using static electricity, which you generated by rubbing one of the special magic crayons back and forth across, say, your mother's draperies, you affixed the vinyl sheet to your actual TV screen. Then you took out your special magic crayons and your special magic eraser and waited for the show to come on.

Here comes Winky Dink, kids! The big lion is after him!! Oh no!!! Winky's trapped!!!! He's got to get across the river but the bridge is out!!! Quick!!! Draw a bridge across the river!!! HURRY!!!!!! You did it!!! Winky made it across the river!!! Now erase the bridge before the lion gets there!!! HURRRRRY!!!!! You did it again, kids!!!! You saved Winky!

I didn't know it then. But Winky Dink was the first interactive TV experience in the history of the world. An idea so creative, it didn't just break ground. It assaulted it. Blew it up. Shook it up and down like Oakland in a 7.8 tremblor.

Winky Dink wasn't a TV show. It was an experience. Even as a kid, I remember thinking, gee, what a cool idea. I wish I had thought of it.

So maybe that's it. Maybe *Winky Dink* spoiled me for all the great advertising that was yet to cross my path in the years ahead. Great as it all was, none of it seemed to get to me like Winky did.[1]

Vocational cues were not the only messages absorbed consciously or subconsciously through the television set. Beth Jarrard offers these reflections among her "TV tales":

I was born in 1950 and virtually grew up watching TV. My mother says I was the quiet, peaceful one among a family of four children, so TV was a great way to amuse myself and to interact with my older brother and sister.

> At a young age, I think I began learning, with the help of TV, some lessons in diversity, tolerance, and acceptance. My mother worked in the family business with my father so we had an African-American cook/housekeeper who also cared for the children. Auntie Julia came to work for my family several months before I was born, and we have always been fast friends. I recall many summer afternoons, when her work was done and her ride home had not arrived, we would watch baseball games together. She would root enthusiastically for many teams, but her favorite players were Willie Mays and Jackie Robinson. This was about the time that I began to ask why Auntie Julia didn't sit down to eat every day with the family. My parents came home in the middle of the day, and we ate our big meal—"dinner" prepared by Julia—at 1 P.M. and a light "supper" at night. Julia would usually do laundry in the basement while we ate and then have her own dinner before clearing the table and doing the dishes. I protested the tradition and began waiting and eating with her, just the two of us.

John Idol notes that adults too learned incidental social lessons from the medium. He thinks of his parents, who grew up in a North Carolina mountain county where there had been no slaveholding in Antebellum times. Thus the Deep Gap area, mainly peopled by white farmers, had always known (and still has) only a very small African-American population. In a region where keeping one's own silent counsel and minding one's own business are prevailing unspoken rules, the possibility of knowing anyone well across the racial line was, like the place itself, remote. After television's long-delayed arrival there, the elder Idols began to watch *Jeopardy, Wheel of Fortune*, and other midday audience participation programs during their lunchtime break from their chores, and they were "astonished to find that the black contestants were as smart as the white ones."

In her pre-Internet era public school years Sherri Butler and her ever-curious brother discovered two means of "extracurricular education." Their parents worked outside the home for long hours, and in that era her wheelchair blocked rather than gave access to the public library, so she and her brother turned to the "approved" help of the family's *World Book Encyclopedia* set and the "scorned" medium of television, each supplementing school work in a different way:

> TV, of course, did not lend itself to [interactive] research. But while the adults shook their heads over figures showing how much time kids spent in front of the TV, questioning its versions of "history" and the value of even its most value-oriented programming, it was the imaginative door that led us to discoveries and interests as rich [as]—richer than—anything we found at school.
>
> School gave us the tools we had to have. TV suggested places we could go with them.
>
> Westerns like *Wagon Train* served up, in an hour of melodrama weekly, the trials of Westward expansion. It didn't pretend to be history, anymore than *Rawhide* was meant to be a faithful representation of a cattle drive. But I wonder how many people were driven to know more after watching.[2]

Roger Rollin had seen little use for television in his childhood and college years, but the ferment of the 1960s meant that his family felt obliged to make room for a TV set, and that choice had consequences:

By 1965 the Rollins were regular TV viewers, although we were still snobs enough to get our news from the daily papers and not the television news shows. And then I received a profound lesson in TV's power. Like everyone else, I had been reading about the ugly situation in Selma, Alabama, where Negroes were demonstrating for their right to vote. I remember reading one night about how a peaceful march of black men, women, and children had been set upon by the police at the Edmund Pettis Bridge and had been brutally beaten with billy clubs and assaulted with horses and dogs. I was angry and despairing, certainly, but when I reached campus the next day I found my colleagues outraged, incensed, and ready to do battle themselves. And then I discovered that they had seen the attack on TV the night before. I went home that night, turned on the television news, and saw the terrible assault replayed. Two days later I was on the road to Selma with nine other colleagues and the next day was marching myself, my spirits raised (in spite of my fear) by Dr. Martin Luther King, Jr.'s address at Brown's Chapel in Selma's Negro district. When the march started we all were on TV, and I credit the presence of television, along with the Alabama National Guard, with helping to prevent violence from being visited upon us marchers.

It is likely to be more than simple nostalgia or an obvious difference between childhood and adult understandings of things that leads pioneering viewers to regard TV's limited-choice and often-clumsy early programming as superior to today's seemingly unlimited offerings. Father knew best, Lassie came home each evening after a full day of rescuing children from drowning or mine collapses, and all seemed well on American television before The Great Unease settled over the country like a pollution cloud. Randy Cox senses that difference:

It was a different world, maybe even a better one. While it is true that we had precious little choice in what to watch, given only three networks and no concept of "pay TV," what was left was sure to be watched by nearly everyone. Even on nights offering little entertainment value, you could rest assured that about a third of everyone in America was tuned in with you. And you could validate that the next day at school or in the office. Life was simple. Go to work or school; come home for afternoon or evening duties or playtime; watch the news over dinner, as this was the only chance to discover what was going on save reading the morning newspaper; and settle in for a night of television with the rest of America.

I remember our first boxy black and white television set, the kind with a pane of glass in front of the picture tube and four short skinny legs supporting all that weight. I also remember countless house calls by the television repairman. I don't even know if that profession exists now. But that behemoth lasted for quite some time. After my family upgraded to a color set, the ancient box remained in the guest room for years, long enough to enter the video gaming revolution. After all, when we purchased that Magnavox Odyssey 2000 device in 1972 which allowed us to play a

Pong clone, we never imagined that anyone would hook up a game system to the set in the family room. How rude! Frolicking with the main source of family viewing seemed as ridiculous as trying to dial a telephone in a moving automobile.

Thus the real-life family in the wood, plastic, and shag-carpeted viewing room liked what it saw of the on-screen family whose living room or chaste bedroom was simulated, usually in a spare style, on the small screen.

In recent decades the phrase "family values" has become a contentious code expression, a broom for sweeping aside those who do not stand within the circle of largely Protestant and implicitly Anglo-Saxon orthodox values. In the first television generation, by contrast, "family values" was a token of praise, as unequivocal as a consumer magazine's stamp of approval for a product or service. Lucy Ricardo's house might have been amusingly messy or her hair mussy, but the *I Love Lucy* domestic brand of comedy was clean, notwithstanding some exaggerated rolling of eyes and smacking of lips when husband Ricky was in view. When too much eye-rolling led, it seemed, to Lucy's pregnancy, CBS made certain that she was invariably described as merely "expecting," not "pregnant," even during the weekly episodes leading up to the widely watched January 19, 1953 "Lucy Goes to the Hospital" outing, which ironically aired on the day that Lucille Ball delivered her real-life second child. Network constraints and the Federal Communications Commission (FCC) list of seven forbidden "bathroom" words, soon to be challenged by George Carlin and others, meant that no parent was ashamed for the children or the visiting grandmother to share the evening program choice. No code letters for sexual content, street and locker-room language, or violence appeared in the corner of the black and white or early color picture. Sexual content was muted, "gosh-darn" sufficed as the expletive for many frustrations, and the violence of *The Untouchables* or *Have Gun, Will Travel* was put down by—or served the purposes of—agents of law and order.

"I go back to the days of *Ozzie and Harriet* and *Mr. Ed*," Elaine Siegel says. "I remember the days when television was just clean viewing instead of every program [being] about sex. Now I often cringe, when I watch television with my 17-year-old son, because of what is being offered. There have been times during our children's early years when we turned the television off (to their dismay) because the programming was not fit for not only my children but for us. I miss the musical shows that Garry Moore and Jimmy Dean had on the air and the comedy of Red Skelton."

Donna M. Williams lists *I Remember Mama, The Loretta Young Show, Howdy Doody*, several Westerns, and *The Arthur Murray Dance Studio* among her early TV favorites. "On occasion," she says, "I would get to spend the night with my godparents, where I was allowed to fall asleep to the TV and got to watch *The Late Show* (good movies back then) and, once in a great while, *The Late, Late Show*." She adds:

The first television we had in our home was in 1956 or 1957; it was an RCA with a 12-inch screen and was given to us by an aunt and uncle, who purchased a new, larger set. My recollection is that from the beginning years through to the early 1980s television programming was much more wholesome and provided [better] family entertainment than it has since that time. It was not so necessary to be selective and to "police" what the "box" brought into your living room. From the mid- to late 1980s forward, I not only became selective about what was watched in my home, but I began watching television less and less, going back to the "basics" of reading, talking, and outdoor activities as alternatives. The language [on television] today is despicable, the action violent to disgusting—it is no wonder our young people do and say the things they do.

Julie A. Dunlap found that a disagreeable change overcame the network sitcoms in the 1980s. "In the shows that I watched while growing up," she explains, "there were intact families and an atmosphere of moral rightness, even innocence [reflecting] the general opinions of the population of that time [and] not apparent in today's shows. Through the '90s and to the present, the same disgust with the network sitcoms has prevailed, and I still spend most of my TV time with PBS, especially *Masterpiece Theatre*, perhaps reflecting my interest in literature, and *Nature*, for the animals. I enjoy some of the animal shows on Discovery and Animal Planet and the mysteries on A&E."

Even in the early days Mary Ann McKenzie's parents insisted that "nothing vulgar" be watched in their home. Jackie Gleason was listed among the vulgar for his general demeanor and for his loudmouthed portrayal of Ralph Kramden in the "Honeymooners" sketches. Soap operas were another lathery subject: "My mother was a worker and a reader, and so she didn't spend much time watching television. There was never a soap opera on at my house. She disliked them. She thought they were tawdry and stupid. And I wasn't. It wasn't as if I wasn't allowed to watch them. I never wanted to, anyway." Her parents took her to the Ringling Brothers circus, to major league baseball games, and to Lincoln Center instead.

More recently, Jennie Wakefield finally yielded to her children's insistence that she sign up for cable television "because their friends were refusing to visit" without access to their shows. Cable reruns have allowed her to reevaluate *Leave It to Beaver, I Love Lucy*, and other favorites of her own childhood, and she injects a note of caution against sweeping approval of them: "I grew up on those shows, as I was born in 1955, and rewatching them on cable now seemed like it would be revisiting the good ol' days—none of that casual sex and violence like we have today. But, while watching with my teenage daughters, I was amazed that the stories are based on pulling the wool over someone's eyes—children covering up something they did, parents covering up what they did, hiding the truth from a family member to save face or to present a certain front to that other person. It's all very funny, of course, but coming to it after several years, and the expectation of "wholesomeness," the comedic foundation of deceit struck me. No sex and violence, but whole shows devoted to getting your way by lying."

Priscilla Kanet was delighted by early television portrayals of Polly Crockett, Wyatt Earp, and Bat Masterson, among others, but a note of disillusionment creeps into her voice when she recalls the domestic comedies of an earlier day: "I think I grew up expecting life to be like TV—*Donna Reed* and *Ozzie and Harriet.* (I had a big crush on Ricky Nelson.) Even when I got married, I thought I needed to be like those women—homemade everything, staying home for kids. To this day, I still think Walter Cronkite was/is the American voice of truth, justice, and . . . oh, that was Superman."

Notes

1. "Advertising: A Case of Winky Dink Envy," *Communication Arts 329* (July 2004): 184.

2. "Accidental Learning," *Herald-Leader* (July 11, 2001): sec. B.

III

What Others Saw

$$--- 11 ---$$

Television Is . . . /Was . . . /Will Be . . .

What *is* television? The first-viewing hundreds, then thousands, then millions have often defined it in terms of viewing companions, of the TV set with a well-remembered quirk, or of their favorite series, memorable specials, or events reporting. It is a wood or plastic box filled with wires and fronted by a cathode tube. On early sets it was the Indian head test pattern and the accompanying steady tone and then, when programming finally started, black and white images seen through snow and noisy interference patterns. On flat screen monitors today it appears to be the animated hieroglyphic wall of the electronic age. Television is webs of networks and affiliates, rusting and sometimes sagging antennas above older homes (or newly mounted antennas for HDTV reception in some areas), monthly cable or satellite subscription bills, a link between viewers and their videotapes, DVDs, or TiVo backlogs. Early in the twentieth century a few engineers, entrepreneurs, and creative dreamers foresaw this medium. Since then some have looked deeply into it, beyond, so to speak, the surface reflection of Narcissus's pool.

Television was a "mistress" to Philo Farnsworth, whose idea for the scanned lines that make up the TV picture was central to the development of the medium. He is said to have confessed to his wife on their honeymoon in 1926, "There's another woman in my life, and her name is Television," and he spent much of the remainder of his life in his laboratories and in the courts, defending his patents.[1]

According to David E. and Marshall Jon Fisher, the word "television" was first used by Constantin Perskyi at the International Electricity Congress at the 1900 Paris Exhibition.[2] They note that this word stuck while other names for the medium fell aside: "visual listening," "audiovision," "telectroscopy," "telephonoscope," "hear-seeing," "raduo," "electric vision," and "electrical telescope."[3] The word was a classicist's nightmare to C. P. Scott, editor of *The Manchester Guardian*, who in 1928 declared his disdain: "Television? The word is half Greek and half Latin. No good will come of it."[4]

As 1920s editorialists, essayists, and educators had hoped that radio would become a noncommercial, culturally rich instrument of "uplift" for all who heard it, so commentators of the 1930s brightly hoped that television would spread ideas and general enlightenment. "Television is here," proclaimed the September 1931 issue of *Radio Digest*, claiming that twenty-two stations were offering programs to viewers of 25,000 receivers in the United States. Not anticipating the effects of the Depression, the magazine predicted, "In another year you will find not less than 100,000 television receiving sets throughout the country."[5] E. B. White, writing cheerfully in his *Harper's* column "One Man's Meat" in 1938, saw the medium as a springboard to metaphysics: "Television will enormously enlarge the eye's range, and, like radio, will advertise the Elsewhere." The television experience would be hyperreal, in his view: "A door closing heard over the air, a face contorted, seen in a panel of light—these will emerge as the real and the true—and when we bang the door of our own cell, or look into another face the impression will be of mere artifice."[6] In his essay White envisioned television as the community center of a wired age, and his often-quoted hope for television sounds like a command too: it "should be our Lyceum, our Chautauqua, our Minsky's, and our Camelot."[7] Nearly a decade later, circa 1947, when television sets were still rare in most citizens' everyday experience, Ron Dubin observed the customers who came into his father's New York City grocery store for a glimpse of the TV screen: "It was such a novelty. People were always commenting on what was said and what they saw. There was as much conversation coming from outside the set as there was coming from the set. Nowadays, people are rather passive when they watch TV. Then, it was a much more active involvement."[8] In an unsigned 1948 *New Yorker* "Talk of the Town" item, E. B White took several steps back from his decade-earlier hopes: "Like radio, television hangs on the questionable theory that whatever happens anywhere should be sensed everywhere. If everyone is going to be able to see everything, in the long run all sights may lose whatever rarity value they once possessed, and it may well turn out that people, being able to see and hear practically everything, will be specially interested in almost nothing."[9]

Certain definitions of (or metaphors describing) television have reached classic, often-quoted status. One cannot consider the early development of the medium for very long without encountering Federal Communications Commission (FCC) Chairman Newton Minow's remarks in a speech to the National Association of Broadcasters in Washington on May 9, 1961, as he urged the industry to scan its own product: "When television is bad, nothing is worse. I invite you to sit down in front of your television set when your station goes on the air . . . and keep your eyes glued to that set until the station signs off. I can assure you that you will observe a vast wasteland." Soon after transcripts of Minow's speech were printed in daily newspapers, a viewer wrote to ask him what day and time "Vast Wasteland" came on the air.[10] Media critic Marshall McLuhan famously declared radio a warm and cuddling medium and television a "cool" one that leaves a glassy glaze of separation between the tube and the viewer. Comedian Bob Hope quipped that television is "that piece of furniture that stares back at you,"[11] while

Fred Allen, whose strong radio career fell victim to television, sourly commented, "Television permits people who haven't anything to do to watch people who can't do anything."[12] In a letter published in the *Times of London* on December 20, 1950, poet-playwright and bank clerk T. S. Eliot found television potentially addictive, noting that his American friends had become "concerned with the television habit" and that they exhibited "only anxiety and apprehension about the social effects of this pastime," especially on the mental, moral, and social development of small children.[13] A major engineering progenitor of television, Vladimir Zworykin, said in an interview on his ninety-second birthday in 1981, "The technique is wonderful. I didn't ever dream it would be so good. . . . It is beyond my expectations. . . . [But as for the programs] I would never let my children come close to the thing. It's awful what they're doing."[14]

For manufacturers' and networks' advertising agencies, television was something else—not the least a subject for straining metaphors and lavish promises. About 1945 a copywriter for DuMont, which both manufactured sets and developed a pioneering (but short-lived) network, borrowed from the circus poster's broadside style: "Coming! Television: The greatest show on earth! Glamorous musicals and the stage's most brilliant dramas! Boxing and ballgames, races and wrestling! Parades, movie premieres, and political conventions . . . running bumper to bumper in the most magnificent pageant ever dreamed!"[15] "Television is the hottest battleground of the entertainment world today," Allan Carpenter concluded in the 1946 *Science Digest* article "Where Is Television?" where he reported that the general public was generally disappointed in the technical limitations of the medium at a time when there were estimated to be 2,500 receivers in New York City and 400 in Chicago.[16] The networks continued to assure would-be advertisers that all would be well—and profitable. A full-page 1951 CBS Radio trade ad claimed, "Television's a wonder-child, and no question about it. Precocious as anything, and big for its age. Almost makes you forget that television's got a big brother that can still lick anybody on the block."[17]

Television thus became the handmaiden to—or the shrieking hag of—commerce. Thomas Hine points out that, although munitions spending and other measures during World War II had seemed to lift the U.S. economy out of the Great Depression, many feared a return to harsh economic conditions in the postwar years. Fortuitously, though, TV sets were first becoming generally available to the public in those years, and in an optimistic 1946 report the U.S. Department of Commerce declared, "Television as an advertising medium will create new desires and needs and will help industry move a far greater volume of goods than ever before."[18] Painting on a somewhat larger historical canvas in his 1967 book *The New Industrial State*, economist John Kenneth Galbraith wrote that when "the masses" had gained enough disposable income, sellers could profit by conveying their messages through a medium that would provide "comprehensive, repetitive, and compelling communication" to "people in all spectrums of intelligence."[19] In Galbraith's view, radio and later television, "in their capacity to hold effortless interest and their accessibility over the entire cultural spectrum" without "any educational qualification" (i.e., literacy), arrived at the ideal time to become "the prime instruments for the management of consumer demand," and he warned those who undervalued the commercial use of broadcast media, "There is an insistent tendency among solemn social scientists to think of any institution which features rhymed and singing commercials, intense and lachrymose voices urging highly improbable enjoyment, caricatures of the human esophagus in normal or impaired operation, and which hints implausibly at opportunities for antiseptic seduction, as inherently trivial. This is a great mistake. The industrial system is profoundly dependent on commercial television and could not exist in its present form without it."[20]

"In the half century since its commercial unveiling," declares the front flap to the dust jacket for David E. and Marshall Jon Fisher's *Tube: The Invention of Television*, "television has become the undisputed master of communications media, revolutionizing the way postwar generations have viewed the world."[21] Historian William Manchester, a confidant of the Kennedys and politically powerful others in the 1960s and 1970s, called television "a great, silent agent of change,"[22] and LeRoy Collins, Florida's governor from 1955 to 1960 and later a voice for civil rights causes, deemed it "the greatest single power in the hands of mortal man."[23] Barbara Lippert, writing of Nickelodeon's Nick at Nite vintage series reruns, called such programs "electronic comfort food" and declared, "The power of old TV is such that it's become the only history we honestly share."[24]

At CBS, newsman Edward R. Murrow and producer Fred W. Friendly were responsible for a number of defining achievements in the 1950s and 1960s: *See It Now, Person-to-Person*, the *CBS Reports* documentary "Harvest of Shame," and the exposure of Sen. Joseph McCarthy's interrogation methods are among the primary credits to one or both of these men. Each attempted to maintain a balanced view of the medium. In characteristically emphatic rhythm, Murrow declared that television "can teach, it can illuminate; yes, it can even inspire. But it can do so only to the extent that humans are determined to use it to those ends. Otherwise, it is merely lights and wires in a box."[25] Murrow had been the host for the first program linking the East and West coasts on November 18, 1951, with pictures of the Brooklyn Bridge and the Golden Gate Bridge on monitors above the smoke curl of his cigarette. His frequent collaborator Fred Friendly later alluded to that beginning of transcontinental television and subsequent milestones: "Television can show you the Atlantic and the Pacific, and television can show you the face of the moon. But it can also show you the face and heart of man. And perhaps what it does best is the latter."[26] However, some disillusionment emerges in his later statements. In a 1966 interview recalled in his *New York Times* obituary thirty-two years later, Friendly said, "TV is bigger than any story it reports. . . . It's the greatest teaching tool since the printing press. It will determine nothing less than what kind of people we are. So if TV exists now only for the sake of a buck, somebody's going to have to change that." Angry in his fight over carrying a live news report from the February 1966 Senate hearings on U.S. military involvement in Vietnam when his CBS bosses insisted on showing the scheduled *I Love Lucy* morning rerun, he declared, "It was . . . a choice between interrupting the morning run of the profit machine—whose only admitted function was to purvey six one-minute commercials every half hour—or electing to make the audience privy to an event of overriding importance taking place in a Senate hearing room at that very moment." (In protest he resigned his CBS position in the next month.)[27] In his foreword to Newton N. Minow, John Bartlow Martin, and Lee M. Mitchell's 1973 Twentieth Century Fund Report *Presidential Television*, Friendly said, "Television is a miracle that American society has never learned to manage and, in its relationship to politics, it has been permitted to run wild. In the case of presidential politics, it has been the means for a corruption of power that has brought the nation to the edge of disgrace. While the Watergate scandal seethed under a Republican administration, it might just as well have plagued a Democratic incumbency," and he complained, " . . . our inability to manage television has allowed the medium to be converted into an electronic throne."[28] Later, CBS news anchor Walter Cronkite complained of the half hour evening newscasts, "Everything is compressed into tiny tablets. You take a little pill of news every day—23 minutes—and that's supposed to be enough."[29]

If TV news pioneers developed some disillusion with television, so did some commissioners of the FCC. One-time agency chairman Mark Fowler dismissed it as "just another appliance, . . . a toaster with pictures."[30] Even more starkly, former commissioner Lee Lovinger said, "Television is the literature of the illiterate, the culture of the lowbrow, the wealth of the poor, the privilege of the underprivileged, the exclusive club of the excluded masses."[31]

Lovinger's characterization is bitterly confirmed in James Baldwin's 1960 essay "Fifth Avenue, Uptown," which describes his return to the Harlem block where he grew up, one side of the street falling into increasing neglect and decay, the other (where Baldwin's family home once stood) dominated by an ill-conceived housing project, a tired place that is becoming reghettoized. Baldwin discovers a new verb as he describes the young denizens who have given up: "They stay at home and watch the TV screen, living on the earnings of their parents, cousins, brothers, or uncles, and only leave the house to go to the movies or to the nearest bar. 'How're you making it?' one may ask, running into them along the block, or in the bar. 'Oh, I'm TV-ing it'; with the saddest, sweetest, most shamefaced of smiles, and from a great distance."[32]

At a gathering of radio and TV news directors in Chicago in 1958, Edward R. Murrow spoke somewhat cynically about sponsors' influence: "Don't be deluded into believing that the titular heads of the networks control what appears on their networks. They all have better taste."[33] Elsewhere in the speech Murrow called television an "intellectual ghetto."[34] Others associated in one way or another with the broadcast industry have grown similarly scornful or cautionary. EMI engineer Bernard Greenhead remembered that after an experimental demonstration of the medium, Sir Isaac Schoenberg told those assembled in the control room, " 'Well, gentlemen, you seem to have perfected the biggest timewaster of all mankind. I hope you use it well.' "[35] In contrast to the abrupt pointedness of Murrow's and

Schoenberg's declarations, bandleader Lawrence Welk made an unwitting remark on the medium's preference for the ordinary when he spoke of the music selections that made up his "champagne music" repertory: "If they can't hum it after we play it, it's not for us."[36] Of another television genre, actor and TV producer Sheldon Leonard said, "Television requires familiar fare because of the conditions under which the material is viewed. It is viewed at home, as a form of relaxation. It is viewed between the upturned, unshod feet and with a can of beer at hand. The homely situation comedy is well-adapted to this kind of viewing."[37] Even Gene Kinsella, the parody TV executive character in the *Murphy Brown* comedy series, came to the media confessional booth in one episode: "Television is nothing but a game of three-card monte designed to keep the natives distracted while the nation goes to hell."[38] Real-life former CBS president Howard Stringer, quoted in Michael Skube's tellingly titled article "If Television's Own Executives Trash It, Who Are We to Disagree?" said it in a different way: "We see a vast media-jaded audience that wanders restlessly from one channel to another in search of that endangered species—originality."[39] Hollywood mogul Darryl Zanuck brushed away the potential competition in 1946: "People will soon get tired of staring at a plywood box every night."[40] Film producer Samuel Goldwyn was equally dismissive in a widely quoted 1956 rhetorical question: "Why should people go out and pay money to see bad films when they can stay at home and see bad television for nothing?" Drama critic Clive Barnes's statement has a kicker: "Television is the first truly democratic medium, the first culture available to everybody and entirely governed by what the people want. The most terrifying thing is what the people want."[41]

"Like it or not," Stephen Seplow and Jonathan Storm have written, "over the past 50 years, TV has become the core of the culture. Television time has become so ingrained that it rivals natural rhythms in daily importance. Does anyone seriously question that evening starts for most people at 8 P.M., the beginning of prime time, rather than when the sun sets in the West?"[42] "Television has given us new ways to measure the passage of moments, years, and our lives," says film historian Tom Shales, and Donna McCrohan, contemplating the presence of the TV set in Archie and Edith Bunker's living room in "*All in the Family*," remarks that " . . . TV *is* a time capsule. . . . "[43] Reporter Nelson Smith found television to be a companion to Mary, a dying woman in an East 20s Manhattan walkup: "She passed the hours in a crushed armchair near the television, which flickered day and night like a coal fire."[44] When Murray Haydon awoke in his hospital bed two days after a heart transplant in 1985, he asked the nurses, "Would you please turn on the television? . . . I would like to see if I'm still alive and how I'm doing."[45] Edward Wakin finds this moment a disturbing illustration of U.S. TV dependency: "It is what Americans rely on to find out about the world. They can't do without it from the time they get up in the morning to the time they go to bed. They can't imagine life without TV telling them what's happening."[46] Americans are not alone in this, however. In 1998, authorities in Potsdam, Germany, found the mummified body of a man, still facing the TV screen, who had died more than four years earlier.[47]

In rural Russia a few years ago a house burned down, and the cow in an attached structure died in the fire, but the family took care to save the TV set.

John Mason Brown has called television "chewing gum for the eyes,"[48] and Eric Bentley has said, "Television is the supermarket of the soul."[49] Newspaper columnist and mystery writer Celestine Sibley recalled of early TV days, "A boy asked me once if it was 'low class' to have a television set in your living room, and I was so bemused by the question I could never think of an answer."[50] Younger-generation columnist Rheta Grimsley Johnson faced the issue of whether or not to have a satellite dish installed, confessing, "I swore never to own one. It's like having a tattoo on your dwelling that says, 'Can't get enough of that idiot box.'"[51]

Some newspaper columnists have scolded the medium for its failures of social responsibility. Max Frankel of *The New York Times* wrote during an election season, "I get 93 channels on my TV screen and not one treats me as a caring citizen. Not even the half-dozen stations that pretend an interest in local news showed any interest in helping me cast an intelligent vote last Tuesday. They got rich from selling political ads but gave the candidates and issues the back of the hand."[52] Syndicated Cleveland writer Dick Feagler spoke, in less than his usual geniality, of the "talk-show, freak-show" patterns of 1990s daytime television: "The TV industry is on the brink of terminal amnesia in matters of taste, restraint, and shame. A sharp blow to the advertising revenue may restore a little responsibility."[53] Ronald K. L. Collins, however, is less hopeful that public demands will alter the medium's essential nature: "Television is at war with our social self: It is inherently alienating; it caters to the solitary life. A room with five people watching TV is not an assembly of people: it is a room of five occupying the same space."[54] A correspondent to the "Dear Abby" advice column, signing herself "Ex-Television Widow," complained, "I was married twice, and both of my husbands tuned me out for that busy box," and Abigail Van Buren replied, "Television can be the subtle thief of precious time—and he or she who falls into the lazy habit of watching just anything that moves is destined to become an intellectual pauper."[55]

On New Year's Day 1997 an anonymous contributor to the *Atlanta Journal-Constitution's* "The Vent" column, probably fresh from sorting out New Year's resolutions, registered a discovery: "I now know why they call television a medium. It isn't well done."[56] Another venter observed later, "We used to have only three or four television channels to pick from, and now there are dozens. Why does it seem that it was easier to find something worth watching back then?"[57] Such sentiments have led in recent years to "Kill your television" bumper stickers and to national TV Turn-Off weeks. Film actor Johnny Depp has encouraged such actions, calling television an enemy to reading and saying, "Books are one of the best things that happened to me. . . . Nobody reads anymore because we've been weaned on the glass breast: TV."[58] According to an excited character in Jan Eliot's syndicated family comic strip "Stone Soup" (where characters often speak in bold lettering with exclamation points), "TV's the new religion. TV replaced free thought. TV replaced real culture. Oh what a crock we bought!"[59] Since 1982 members of the Society for the Eradication of Television have carried wallet cards that read, "I do not own a television."[60]

In both an article for *The Saturday Evening Post*[61] and in a subsequent book title Marie Winn has called television "the plug-in drug."[62] Jean Lotus has admitted "that TV was her 'drug of choice' as 'a nerdy junior high student,'" but she later called it "a sewer pipe to the soul" and directed her energies toward publishing the newsletter *"The White Dot"* (named for the lingering point of light near the center of the screen when a TV set is turned off),[63] which achieved thirteen issues before she discontinued publication in 2000 because of "my own exhaustion" and the immediate needs of her young children. Nonetheless, she observes with pride that her pioneering Internet advocacy has directly or indirectly encouraged the development of thirty current anti-television Web sites in a day when 2 percent of the U.S. population does not own a TV set.[64] In her advocacy of a TV-free or at least a TV-rationed lifestyle she has said, "TV should be like a spigot, turned on and off. It shouldn't be a constant flood.... TV takes you away from friends and family and sells family feeling back to you as you sit there alone. It's so depressing. I think it says so much about us that one of the most popular shows is called '*Friends*,' but most people watch it alone."[65]

During the 1998 TV Turn-Off week, *Miami Herald* columnist Leonard Pitts argued for viewer self-discipline:

> ... television is, for better or worse, part of the connective fiber of our society. As the nation frays into ever-smaller subsets of culture, gender, class, and politics, it's one of the few forces short of a righteous war still able to bring millions of us together on the same page at the same time.
>
> Television is a seducer, there's no denying this. A thief of time, as well. But it's also entertainer and edifier.
>
> Turn it off for a week? Overkill, if you ask me. Turn it off for a few hours instead.
>
> Because it's not television that makes us soft and stupid. It's not knowing how to not watch. It's always needing sound and pictures to fill silences and blanks.[66]

Todd Gitlin has called television both "a panderer" and "the storyteller for the American tribe,"[67] and his book *The Twilight of Common Dreams: Why America Is Racked by Culture Wars* offers an insightful extended definition of early television:

> The small screen was far more than a machine for perpetual entertainment. Television was an amusement bank, a national bulletin board, a repertory of images, an engine for ideas, a classification index, a faithful pet, and a tranquilizer. Perhaps most of all, the television networks were crucial dispensers of America's master idea of itself. The dream of normality was incarnate in *The Adventures of Ozzie & Harriet, Father Knows Best, The Donna Reed Show*, and comparable rituals of cheerfully intact white families living the good life within their white picket fences and solving their silly problems. The boundaries of morality were marked by the

likes of *Dragnet* and *The F.B.I.* policing transgressions. *The Lone Ranger* and *Gun-smoke* gave the moral law roots in a mythological history. For exotica, Americans could turn to *Disneyland*, hosted by folksy Walt Disney of Kansas City, certifying that we were a fun-loving, innocent people who deserved to be on top of the world.

All in all, television flattered Americans that they had never had it so good—and that they had a common destiny. According to programs and commercials alike, they had no regions, no accents, no divergent histories. Television was a school for manners, mores, and styles; for repertories of speech and feeling; even for personality. Television helped teach Americans how to talk, look, and behave—which meant, in some measure, teaching them how they should think, how they should feel, and how, perchance, they should dream. And so, while certifying the high-consumption way of life, network television also served as an instrument for the nationalization of culture—and in the end, therefore, helped further a certain bland tolerance while eroding enthocentrism and other forms of parochialism.[68]

More briefly, the dust jacket front flap to Donna McCrohan's *Prime Time, Our Time* gives her view: "Television is not a medium for the avant garde; it is a medium that defines our current tastes, values, social concerns—all within what is acceptable to the majority of its audience."[69] In the Introduction to a book of Lloyd DeGrane's photographs of people watching television in a variety of settings, William Brashler offers a similar conclusion: "TV has become the pastime of our culture—an all-pervasive national pastime, something just about everybody grew up with. Generally speaking, you have to be at least fifty years old to have grown up in a house without a tube."[70] A reviewer in the professional library journal *Choice* says simply that television is "society's most symptomatic medium."[71] Thus CBS president Frank Stanton proved to be both planner and prophet when he said in 1948, "Television, like radio, should be a medium for the majority of Americans, not for any small or special groups; therefore its programming should be largely patterned for what these majority audiences like and want."[72]

Notes

1. Quoted in "*Modern Marvels:* Television: Window to the World," The History Channel, 1999; repeated September 3, 2003. On Farnsworth's marriage and work, see also David E. and Marshall Jon Fisher, *Tube: The Invention of Television* (Washington, DC: Counterpoint, 1996), 134.

2. Fisher and Fisher, *Tube: The Invention of Television*, 357.

3. Ibid., 8, 29.

4. Ibid., 355.

5. Reprinted as filler item "Television," *NARA News*, 25(3) (1997): 33.

6. E. B. White, "One Man's Meat," *Harper's* 177 (October 1938): 553.

7. Quoted in Edward Oxford, "TV's Wonder Years," *American History* 30(6) (1996): 26.

8. Quoted in Jeff Kisseloff, *The Box: An Oral History of Television, 1920–1961* (New York: Viking, 1995), 126.

9. "Talk of the Town: Television," *The New Yorker* (December 4, 1948): 25; authorship confirmed and item reprinted in E. B. White, *Writings from The New Yorker, 1927–1976*, ed. Rebecca M. Dale (New York: HarperCollins, 1990), 175.

10. "Television: Window to the World," *Modern Marvels*, The History Channel, first broadcast 1999, repeated September 3, 2003.

11. Quoted in Oxford, "TV's Wonder Years," 26.

12. Ibid., 69.

13. "The Television Habit," *Times* (London) (December 20, 1950): 7.

14. Quoted in Albin Krebs and Robert McG. Thomas, Jr., "Notes on People: TV Turns Off Its Father," *New York Times* (July 31, 1981): sec. B.

15. Quoted in Kisseloff, *The Box: An Oral History of Television, 1920–1961*, 100.

16. Allan Carpenter, "Where Is Television?" *Science Digest* (December 1946): 54.

17. "Television's Big Brother," *New York Times* (June 18, 1951): sec. C; rpt. Cipe Pineles Golden, Kurt Weihs, and Robert Strunsky, eds., *The Visual Craft of William Golden* (New York: George Braziller, 1962), 95.

18. Thomas Hine "Commercials in Flux: What's around the Corner for Ring around the Collar?" *New York Times* (May 30, 2004): sec. D.

19. John Kenneth Galbraith, *The New Industrial State* (Boston, MA: Houghton Mifflin, 1967), 207.

20. Ibid., 208.

21. Fisher and Fisher, *Tube: The Invention of Television*, dust jacket.

22. Quoted in Edward Wakin, *How TV Changed America's Mind* (New York: Lothrop, Lee & Shepard Books, 1996), 15.

23. Quoted in William Brashler, Introduction to Lloyd DeGrane, *Tuned In: Television in American Life* (Urbana and Chicago: University of Illinois Press, 1991), n. p.

24. Barbara Lippert, "The Image: About Last Nite," *New York* (April 21, 1997): 20.

25. Quoted in Oxford, "TV's Wonder Years," 69.

26. Ibid., 68.

27. "Fred W. Friendly, CBS Executive and Pioneer in TV News Coverage, Dies at 82," *New York Times* (March 5, 1998): sec. B.

28. Fred Friendly, Foreword to Newton N. Minow, John Bartlow Martin, and Lee M. Mitchell's 1973 Twentieth Century Fund Report, *Presidential Television* (New York: Basic Books, 1973), vii.

29. "Update: Walter Cronkite Still at the Helm," *Newsweek* (December 5, 1983): 28.

30. *Current Biography Yearbook 1986* (New York: H.W. Wilson Co., 1987), 152.

31. Joel Brinkley, "Defining TV's and Computers for a Future of High Definition," *New York Times* (December 2, 1996): sec. C.

32. James Baldwin, *Nobody Knows My Name: More Notes of a Native Son* (New York: Dial Press, 1961), 59.

33. Edward R. Murrow, "Views and Reviews: A Broadcaster Talks to His Colleagues," *The Reporter* (November 13, 1958): 35.

34. Ibid., 20.

35. Quoted in Michael Winship, *Television* (New York: Random House, 1988), ix.

36. Quoted in Bernice McGeehan, "Lawrence Welk: Champagne and Grace Notes," *Saturday Evening Post* (March 1980): 52.

37. Quoted in Frank Bruni, "Sheldon Leonard, 89, Actor and Producer of TV Hits, Dies," *New York Times* (January 13, 1997): sec. A.

38. Quoted in Phil Kloer, "Channel Surfer: Murphy Often Told Us What We Needed to Hear," *Atlanta Journal-Constitution* (May 18, 1998): sec. D.

39. *Atlanta Journal-Constitution* (September 16, 1997): sec. B.

40. Quoted in Philip Collins, *The Golden Age of Televisions* (Santa Monica and Los Angeles, CA: General Publishing Group, 1997), 10.

41. Quoted in Oxford, "TV's Wonder Years," 69.

42. Stephen Seplow and Jonathan Storm, "The Little Box at the Core of Our Culture," *Atlanta Journal-Constitution* (January 11, 1998): sec. H.

43. Donna McCrohan, *Prime Time, Our Time: America's Life and Times Through the Prism of Television* (Rocklin, CA: Prima Publishing & Communications, 1990), 224.

44. Nelson Smith, "Lives: Secret Admirers," *New York Times Magazine* (August 23, 1998): 64.

45. Quoted in Claudia Wallis, "The Artificial Heart: Act III," *Time* (March 4, 1985): 69.

46. Wakin, *How TV Changed America's Mind*, 15.

47. "Mummified Body of German Found in Front of TV Years Later," *Greenville News* (June 27, 1998): sec. A.

48. Quoted in DeGrane, *Tuned In: Television in American Life*, n. p.

49. Quoted in McCrohan, *Prime Time, Our Time*, 1.

50. Celestine Sibley, "A Class Act Should Decide Person's Status," *Atlanta Journal-Constitution* (April 5, 1999): sec. C.

51. Rheta Grimsley Johnson, "Pizza on Roof Could Use Some Pepperoni," *Atlanta Journal-Constitution* (September 18, 1998): sec. C.

52. Max Frankel, "Word & Image: Pay-Pay-Pay-Per-View," *New York Times Magazine* (November 4, 1998): 36.

53. Dick Feagler, "Blow to Advertising Revenue May Restore Responsibility to TV," *Greenville News* (November 17, 1995): sec. A.

54. Ronald K. L. Collins "TV Subverts the First Amendment," *New York Times* (September 19, 1987): sec. A.

55. *Atlanta Journal-Constitution* (February 7, 1997): sec. H.

56. *Atlanta Journal-Constitution* (January 1, 1997): sec. A.

57. *Atlanta Journal-Constitution* (May 5, 1999): sec. D.

58. Quoted in Lorrie Lynch, "Who's News," *USA Weekend* (July 31–August 2, 1998): 2.

59. Jan Eliot, "Stone Soup" (comic strip), *United Press Syndicate* (May 2, 1999).

60. Jean Lotus, telephone interview (June 9, 2004).

61. Marie Winn, "The Plug-In Drug," *Saturday Evening Post* (November 1977): 40–41, 90–91.

62. Marie Winn, *The Plug-In Drug* (New York: Viking, 1977).

63. Quoted in Cathy Hainer, "Turning Off the TV Gets Woman Plugged into Living," *Greenville News* (July 4, 1997): sec. B.

64. Jean Lotus, telephone interview (June 9, 2004).

65. Quoted in Hainer, "Turning Off the TV Gets Woman Plugged into Living," sec. B.

66. Leonard Pitts, "TV Keeps Us on the Same Wavelength," *Atlanta Journal-Constitution* (April 30, 1998): sec. A.

67. Quoted in Seplow and Storm, "The Little Box at the Core of Our Culture," sec. H.

68. Todd Gitlin, *The Twilight of Common Dreams: Why America Is Racked by Culture Wars* (New York: Metropolitan Books/Henry Holt, 1995), 64–65.

69. McCrohan, *Prime Time, Our Time*, dust jacket.

70. DeGrane, *Tuned In: Television in American Life*, n. p.

71. M. Yacowar, review of David Marc, *Bonfire of the Humanities: Television, Subliteracy, and Long-Term Memory Loss. Choice* 33 (1995): 456.

72. Quoted in Holcomb B. Noble, "Frank Stanton, Who Guided CBS into the Television Era, Dies at 98," *New York Times* (December 26, 2006): sec. C.

$$\text{—— } 12 \text{ ——}$$

TV and Its Viewers in Other Media

Television is a popular art with some high art aspirations. The video medium, itself an electronic lens, has been examined under a variety of other lenses, as a survey of TV representations in the visual arts, fiction, essays, and autobiographies will show.

TV in Visual Media

Some of the most memorable portrayals of early television appeared in CBS institutional advertising. William Golden, the network's Creative Director of Advertising and Sales Promotion, commissioned illustrations from artists such as David Stone Martin and Ben Shahn, whose work was familiar to museumgoers and collectors. Especially notable among Shahn's contributions is a typically jagged-line drawing of many clustered TV antennae for a 1955 double-page newspaper advertisement titled "Harvest," the text of which begins, "Each year America's rooftops yield a new harvest—a vast aluminum garden spreading increasingly over the face of the nation. The past season produced a bumper crop on all counts: $3\frac{1}{2}$ million new antennas bringing the total number of television homes to 34,567,000. The average family spent more time watching its screen than ever—*5 hours and 20 minutes a day*."[1] The antennae drawing was also used in a "Harvest" promotional folder whose cover contained a parallel illustration, "Wheat Field," which reinforced the suggestions that the burgeoning growth of rooftop antennae was a natural and repeatable process and that TV programming was as rich and nutritious as the matured stalks of grain shown.[2]

William Golden asked Paul Strand, who had cofounded The Photo League with Berenice Abbott in 1936, to take a photograph of antennae against the Manhattan skyline, and for about six weeks Strand and a friend lugged a heavy 8×10 view camera to the roof of one building after another, first on the East Side, then on the

West, until the light, the clouds, and the framing seemed right.[3] The resulting image appeared in a CBS double-page advertisement in the September 1949 issue of *Fortune*, with the prophetic statement "It is now tomorrow . . . " superimposed on the photograph's upper right edge, leading the reader's eye to the right page text asserting that "Tomorrow" means "television" for wise advertising directors, who are invited to "[l]ook closely" at the "new horizon" where clustered antennae represent "not the shapes of things to come, but of things already here. For in Autumn 1949, television in its *full* proportions" is clearly visible " . . . creating new patterns in the basic habits of Americans," including "the habit of tuning to CBS Television."[4]

Norman Rockwell's cover paintings for *The Saturday Evening Post* have sometimes been called "sentimental" and "nostalgic," but his large canvas "The New Television Antenna," used for the November 5, 1949 *Post* cover, certifies that television has arrived in small city America. The young antenna installer is shown straddling the sharp-peaked roof and turning the mounting pole with his right hand, while he leans forward to hear the directions of a suspended older man who pops out of the top floor window to signal the moment when the TV screen inside has reached its peak of sharpness (or least snowiness). The Victorian house shows signs of decay—bricks fallen from the chimney, spindles missing from the captain's wheel decoration just below the roof's peak—but the antenna and its pole are shiny and sturdy. As Anne Knutson observes in the catalogue for a traveling exhibition of Rockwell's paintings, this picture "exemplifies how the *Post* introduced and humored its readers into modern technological society. . . . The architecture wraps the television into something worn, weary, and familiar, blunting the newness of the technology. Rockwell deliberately used the familiar pyramidal composition of traditional religious paintings to construct his house, with the antenna taking the place of the cross at the top."[5] The foregrounded antenna is also echoed by the distant church spires, as if to suggest that it too draws inspired or inspiring messages from the sky.

Far from the traditional subjects and painting technique of Norman Rockwell were the world of performance art and other manifestations of avant-gardism in the 1960s and 1970s. Nam June Paik, whom Bruce Kurtz called "The Zen Master of Video" in the title of an arts magazine profile,[6] used television sets, or parts of them, to make a comment on the nature of the medium. His "T.V. Buddha" trained a closed-circuit camera on a two-centuries-old Japanese sculpture to contemplate, as Kurtz says, "the time dimensions of his medium," as if to say that viewers might find greater meaning in such a static contemplation than in watching all the episodes of *Days of Our Lives* run end to end. "Fish T.V.," made in 1979, replaced the picture tube with a bowl of live goldfish, perhaps an ironic comment on NBC's "In Living Color" designation of its early color telecasts. A composer as well as a visual artist, Paik collaborated with American Symphony Orchestra cellist Charlotte Moorman in a series of events. Bruce Kurtz notes, " . . . in *T.V. Bra for Living Sculpture*, first performed in 1969, Moorman plays her cello while wearing two tiny plastic-encased television sets as a bra; each of the receivers shows live images of her performance, modulated by the changing sounds of the cello. A play on the cliché 'boob tube,' the *T.V. Bra* also calls to mind the intimacy and tactility of America's nourishing font and electronic baby-sitter, the television."[7]

In 1991, Chicago-based photographer Lloyd DeGrane published a collection of thirty-six images showing Americans watching their television sets, either alone, with friends or family, or, in one picture, with a pet llama, which seems nearly as attentive to the large-screen picture as the two human beings are. In the Preface to *Tuned In: Television in American Life*, DeGrane tells of the idea that struck him as he was delivering phone books in the city's South Side. Knocking "on a door in a darkened hallway," he was invited into a nearly bare room where "silhouettes of people" were "illuminated by the flickering light of a television set.... At that moment I realized I had crossed into another world. In the early 1970s, watching television was like a prehistoric clan gathering where the TV was the focus and the people were huddled around a cave fire." Watching the closing minutes of a Godzilla film with the absorbed family, DeGrane was then asked what he wanted.[8] Despite the coolness of that reception, William Brashler comments in the Introduction to DeGrane's collection, "The TV room is the warmest, most central room in the house, and often the messiest, because people eat, smoke, lounge, and scratch themselves, each other, and the dog while they watch TV...." Thus America has embraced TV, which "is more than forty years old now. Middle-aged. It is in constant flux, and perhaps growing a little flabby in the middle. It is also becoming, if you can imagine, *smaller* than it once was."[9]

Beverly Hills postal worker Mark Bennett spent twenty years drawing "fantasy blueprints" detailing the habitations shown in favorite TV series. Drawing on memories from an isolated TV-watching childhood in Chattanooga, he fleshed out floor plans and furniture arrangements in settings from *I Love Lucy*, the Clampetts' new Hollywood home in *The Beverly Hillbillies*, the 4077 M*A*S*H compound, and all of Gilligan's Island, and in 1996 TV Books published a collection of forty blueprints showing the settings for thirty-four prime time series.[10] An *Art in America* critic, reviewing an exhibition of these blueprints in 1998, said that Bennett's "campy humor mingles with a frustrated childish desire for wish fulfillment: Why isn't my family like those on TV?... His blueprints are achingly familiar to those who grew up in the early days of television."[11]

TV in Cartoons and Comic Strips

A four-panel comic strip entitled "Now That We Have Television," printed in 1929 as an advertisement for a lunchroom, anticipates a world of two-way television viewing: a husband telephones his "wifey" to say that he will be late for dinner, she presses a button on a box marked "Television," and a telephonic picture of the husband embracing and kissing his secretary appears on an angled glass screen. When the errant husband arrives home in the last panel, the wife brings a rolling pin down on his head and warns him, "This time you forgot the *television*, but tomorrow don't forget to *tell your stenographer* to look for another *job!*"[12]

Newspaper panel cartoonists saw TV as a programmed medium rather than a surveillance device, but their anticipations were seasoned by their experience of commercial radio. A November 22, 1934 *Chicago Tribune* panel by Gaar Williams shows a radio blasting its "mushy songs an' jokes" into the face of its owner, who naively hopes for something better from the next medium: "Well—maybe television will help a little . . . ." From the late 1930s until the early 1950s syndicated cartoonist H. T. Webster took a weekly swipe at radio under the recurring title "The Unseen Audience," and in the postwar era television saw its share of abuse. In an October 26, 1949 panel Webster features the flustered owner of a new TV set that provides his guests nothing but interference patterns, which they call "the chain lightning effect," "the spaghetti pattern," and "swell wallpaper designs." H. T. Webster's lean cartooning style strongly contrasts that of George Lichtenstein ("Lichty"), whose loosely drawn "Grin and Bear It" cartoons teased the broadcast media from 1932 to 1974. In his July 13, 1961 panel Lichty shows a broken window and two hiding boys, one saying to the other, "My folks think we see too much violence on television. . . . They just won't give us credit for having ideas of our own, Otis!" Jim Berry, in the May 9, 1967 Newspaper Enterprise Association panel "Berry's World," depicts a young couple hanging up the coats of their guests, a dignified older husband and wife. The young woman gleefully whispers to her husband, "Oh, goodie! They happened to drop in on us [while] the TV set is out being repaired—they'll probably think we're *intellectuals!*" In 1954, *Philadelphia Bulletin* staff artist Bil Keane (better known after 1960 for "The Family Circus") launched the gag-a-day TV-listings-page panel "Channel Chuckles," which would be syndicated by the *Des Moines Register* for twenty-two years. It included the character Aunt Tenna, whose hairstyle rose to a t-shape like that of the most modest kind of aerial. On a Web site displaying samples of his early work, Bil Keane says, "The TV repairman was at our house so much I thought he was part of the family."[13]

For years Hank Ketcham's never-growing Dennis the Menace has relied on "telebision" to keep him abreast of the exploits of tassel-shirted Cowboy Bob, and in the June 4, 2001 panel Dennis asks his irritable neighbor Mr. Wilson to tell younger pal Joey "how you used to hafta get up an' walk to the TV to change the channel." When the father in Rick Detoiro's March 5, 1999 "One Big Happy" comic strip brings an old black and white set down from the attic, the two children find its picture "weird" and "creepy," and daughter Ruthie declares, "Oh, it hurts my eyes." In "Blondie," Dagwood Bumstead has a distinctively contented way of stretching his legs as he gazes at the TV screen, while some episodes of the contemporary Fox animated series *The Simpsons* begin with a view of the Simpsons watching television, passing its product through those silly heads.

From its determinedly antiestablishment beginnings in 1952, *Mad Magazine* found targets in TV genres, characters, shows, and sponsors: for instance, mild Howdy Doody was transformed into the menacing horror movie figure "Howdy Doodit!" Frank Jacobs wrote a series of mock obituaries for TV characters, including gentle and reflective Jon-Boy Walton, who became a homicidal maniac and "turned the revolver on himself" after slaying all the other members of the Walton's Mountain household.[14] In her celebration of the publication's history, Maria Reidelbach explains that the *Mad* staff found early television an attractive target not only because it was "insipid" but also because it was manipulative: "Television is taken for granted as entertainment, but most of us fail to remember, or realize, that the real product of television is not the programming—the product is *us*, the viewers, and we are being sold to the actual consumers of television, the advertisers."[15]

Robert Crumb brought the graphic critiquing of television into the fleshy and designedly shabby underground comix of the 1970s and later. In an autobiographical sequence included in the retrospective *The R. Crumb Coffee Table Art Book*, he places his own introduction to TV in the context of his family's preference for suffocating conventionality. Wanting to be "modern people," they bought a suburban tract home, and the installers lugged the first available console TV set into it. In two panels young Crumb sings a dozen of the commercial jingles that he learned while sitting in his designated place as the family watched the tube, and in the next panel he suggests that attending Sunday church services was little more satisfying because, like TV-watching at home, it was steeped in confining ritual.[16]

TV in Films

Like 1920s cartoonists and the artists who painted covers for Hugo Gernsback's many science fiction periodicals, filmmakers of that era misperceived the major direction of TV's evolution. In their well-illustrated catalogue for the Smithsonian Institution traveling exhibition "Yesterday's Tomorrows: Past Visions of the American Future," Joseph J. Corn and Brian Horrigan characterize Bela Lugosi's *Murder by Television* as "a murky and altogether dreadful movie of 1935,"[17] and they show a picture-phone being used in a still from the same year's British science fiction film *Transatlantic Tunnel*. Leaping past an early TV image of Frankie Thomas costumed as Tom Corbett, Space Cadet,[18] they tellingly include a lobby card for the 1956 film adaptation of George Orwell's *1984*. Here an authoritarian observer peers into a telescreen image of lovers kissing, and large type screams the privacy question, "WILL ECSTASY BE A CRIME...in the terrifying world of the future?"[19]

Television's relationships to privacy issues and social control are central themes in two 1998 films. In *The Truman Show*, directed by Peter Weir, Jim Carrey stars as Truman Burbank, who discovers that his "ideal" Florida development is not the place it seems. From birth this "true man" has been the unwitting star living in the false world of an elaborate setting for a TV show, but a Klieg light falling from the "sky" prompts his reassessment of where, in fact, he is. Scriptwriter Andrew Niccol told an interviewer that the film's core idea is the media's power to warp reality: "From Truman's point of view, I guess I like the idea of questioning the authenticity of our lives. I would love for people coming out of the theater to look twice at whoever they're with."[20] Television functions as a time machine in

Pleasantville: through the machinations of a mysterious TV repairman played by Don Knotts, a contemporary living-color high school student, fond of 1950s TV domestic comedy reruns, is transported with his sister into the black-and-white world of those shows and finds life there to be disturbingly constrained. Parents sleep always in separate twin beds, the dialogue is early sitcom formulaic, and the bland Pleasantville weather forecaster invariably promises "high 72, low 72." The *Father Knows Best* and *Make Room for Daddy* value system is shown to be devoid of feeling beyond a surface cheerfulness of family togetherness, and *Time* reviewer Richard Corliss thus called the film a "parable of repression and awakening."[21] Offering a contrarian view, columnist Russell Baker wondered if TV is better for having rejected 1950s values: "The real joke here is on the 1990s, which have no sitcom to compare with the Fifties' *I Love Lucy* and no TV humor at all to compare with the Fifties' Sid Caesar shows. Check out *Saturday Night Live* or *Politically Incorrect* to see how letting it all hang out in Nineties style has done in TV humor."[22]

Paddy Chayefsky's *Network* (1976) is *The Mary Tyler Moore Show* soaked in brine. In the 1970–1977 CBS comedy, small-scale power games are played in the newsroom of ratings-starved WJM, where news anchor Ted Baxter (Ted Knight) often stumbles on the puffy self-importance that masks a chronic insecurity, while Sue Ann Nivens contrives in her smilingly teasing way to advance the interests of her homemaking show and her amatory needs. While the incompetent Ted Baxter is the only staff member not fired after a change of station ownership in the final episode of *Mary Tyler Moore*, network anchor Howard Beale (Peter Finch) responds to his firing in a drunken rage in the early scenes of *Network*. In one of his last on-air appearances Beale demands that his viewers rush to their windows, rush to the telegraph office, send letters to Washington and demand accountability in the broadcast industry by proclaiming, "I'm mad as hell, and I won't take it any more." As the news department falls under the entertainment division, program director Diana Christensen (Faye Dunaway), an unlaughing Sue Ann Nivens type, contrives to bring back the unbalanced Beale as the center of *The Network News Hour*, an intellectual freak show meant more to entertain and amaze than to inform. Collaborating in her scheme and having a brief love affair with her, United Broadcasting Systems news director Max Schumacher (William Holden) makes his "big speech" as he finally rejects Diana and her manipulations: "You're television incarnate, Diana—indifferent to suffering, insensitive to joy. All of life is reduced to the common rubble of banality. War, murder, death [are] all the same to you as bottles of beer, and the daily business of life is a corrupt comedy. You even shatter the sensations of time and space into split seconds, instant replays." He sees the end of their affair as a conventional "happy ending" in a TV melodrama: "... wayward husband comes to his senses, return to his wife.... Heartless young woman left alone in her arctic desolation. Music up with a swell. Final commercial. And here are a few scenes from next week's show."[23] Consonant with Schumacher's characterization of the medium as an indifferent and cold parade of images, the final scene, in which ranting Howard

Beale is assassinated, is set against a wall of monitors competing with each other in a multiform display of TV business-as-usual: glimpses of typical programs are seen, freckled Mikey is there in a vintage Life cereal commercial, and the Canada Dry label fills one of the screens. Beale is simply sucked into the ongoing TV synthesis, grist for that mill.[24]

After the theatrical release of *Network*, some reviewers mumbled that scriptwriter Chayefsky had turned upon the institution that had brought him fame in live 1950s productions of *Marty*, *The Bachelor Party*, and other television plays. However, Rob Nixon, in a background essay on the "Turner Classic Movies This Month" Web page, reports that, despite those early successes, Chayefsky "was always wary of the medium and its potentially negative influence; his son later spoke of how Chavevsky [sic.] restricted him from watching TV and constantly railed against the 'junk' he felt was being shown.... And by the 1970s, the changes in television had given him even more reason to lament its affects [sic.] on society. 'It's all madness,' he said. 'People are instant now. Thanks to TV we have all developed a ten-minute concentration span.'"[25] Roger Ebert, in one of his periodic retrospective reviews for *The Chicago Sun-Times*, observes that this film "caused a sensation" when it was released, but "[s]een a quarter-century later, it is like prophecy. When Chayefsky created Howard Beale, could he have imagined Jerry Springer, Howard Stern, and the World Wrestling Federation?"[26]

TV in Novels

Within the last two decades, novels about television viewers have portrayed zany characters whose compulsions and obsessions for TV and its programs bring them to states of surreality.

In the opening paragraphs of Wally Lamb's 1992 seriocomic *She's Come Undone*, the protagonist, Dolores, watches "two delivery men carry our brand-new television set up the steps" of her family's rented house. After the deliverers "do things to the set," one of them urges the girl to try its plastic knobs:[27] "Now, I hear and feel the machine snap on. There's a hissing sound, voices inside the box. 'Dolores, look!' my mother says. A star appears at the center of the green glass face. It grows outward and becomes two women at a kitchen table, the owners of the voices. I begin to cry. Who shrank these women? Are they alive? Real? It's 1956; I'm four years old. This isn't what I expected." Like Eve encountering the snake, she grows accustomed to her altered perception, applauding with the studio audience when the woman "with the saddest life" tells the story that wins "the loudest applause" among *Queen for a Day* contestants. "I made my hands sting for these women," she says.[28] By the time she goes away to college, TV-watching has become a compulsive behavior; she stuffs herself on TV as she stuffs her body with food, and it drives her to a private mental hospital where her undoing can only be slowed, not cured.

In *Memories of My Father Watching TV*, published in 1998, Curtis White describes another watch-at-all-costs viewer: "The defining memory of my father is of a man . . . reclined on a dingy couch watching TV" and "ignoring the chaos" of the narrator fighting with his sisters. The younger girl, Winny, became "so desperate to be seen" and recognized by her father that she began "passing rapidly back and forth before the TV, through my father's view," to no avail.[29] "My father," the narrator later says, "would watch *Combat* obsessively if for no other reason than that he had an idolatrous relationship with the sound of guns,"[30] and the tale elaborates an irony suggested in the book's Prologue: "And the TV? It is an oracle. It is speaking to us. It has something very important to say. Tonight it is presenting . . . Our Shows."[31]

The everyday unreality of television is well suggested in Marjorie Klein's 2000 novel *Test Pattern* and its well-chosen epigraph from ABC's 1963–1967 series *The Outer Limits:* "There is nothing wrong with your television set. Do not attempt to adjust the picture. We are controlling transmission. We will control the horizontal. We will control the vertical. For the next hour, sit quietly and we will control all you see and hear. You are about to experience the awe and mystery that lead you from the inner mind to . . . *the outer limits.*"[32] Like *She's Come Undone*, this novel begins with the introduction of a TV set into a 1950s blue-collar household. Unable to sleep after the excitement of seeing her first programs, Cassie rises from her bed and creeps downstairs to touch the on-off knob: "My hand seems to belong to someone else. My fingers and toes prickle like they've fallen asleep, and my whole body freckles with electricity." When she turns it on, the test pattern begins to whirl, and thus she begins to see patterns and, later, programs that no other member of the household can see. It is as if the TV were more than fulfilling her mother's earlier wish: "The idea of owning one had once seemed a fantasy, like owning a box full of stars."[33] Cassie's eerie dialogues with the visions on the screen underscore the detachment from reality that might creep into a family's life with the arrival of the box.

The dust jacket of John Grisham's novel *A Painted House* calls it "a story inspired by his own childhood in rural Arkansas." Luke Chandler is accustomed to hearing St. Louis Cardinals games on the radio at his country home, but on a trip into town with his father, Luke is invited to see a World Series game on the new television of Pop and Pearl Watson:

> My heart froze; my mouth dropped open. I was too stunned to move. Three feet away was a small screen with lines dancing across it. It was in the center of a dark, wooden cabinet with the word Motorola scripted in chrome just under a row of knobs. Pop turned one of the knobs, and suddenly we heard the scratchy voice of an announcer describing the ground ball to the shortstop. Then Pop turned two knobs at once, and the picture became clear.
>
> It was a baseball game. Live from Yankee Stadium, and we were watching it in Black Oak, Arkansas![34]

Back home at suppertime, Luke "talked nonstop about the game and the commercials and everything I'd seen on Pop and Pearl's television. Modern America was slowly invading rural Arkansas."[35]

Tony Earley is coy about the amount of autobiographical material in the "essays" that comprise his 2001 book *Somehow Form a Family*, teasingly subtitled *Stories That Are Mostly True*. The first chapter opens by linking the young narrator with his TV doppelganger: "In July 1969, I looked a lot like Opie in the second or third season of *The Andy Griffith Show*."[36] He too lived at the edge of a small North Carolina town, and the family TV set was a measure of status: "I knew we were poor only because our television was black and white. It was an old Admiral, built in the 1950s, with brass knobs the size of baseballs." Reception was a great problem ("The picture flipped up and down") until the narrator and his sister got off the school bus one day and sensed something strange about their home:

Shelley and I spotted what was wrong at the same time. A giant television antenna had attached itself to the roof of our house. It was shiny and tall as a young tree. It looked dangerous, as if it would bite, like a praying mantis. The antenna slowly began to turn, as if it had noticed us. Shelley and I looked quickly at each other, our mouths wide open, and then back to the antenna. We spirited up the driveway.

In the living room, on the spot occupied by the Admiral that morning, sat a magnificent new color TV, a Zenith, with a twenty-one-inch screen. Its cabinet was made of real wood. *Gomer Pyle, U.S.M.C.* was on. I will never forget that. Gomer Pyle and Sergeant Carter were the first two people I ever saw on a color television. The olive green and khaki of their uniforms was dazzling. Above them was the blue sky of California. The sky in California seemed bluer than the sky in North Carolina.

When Shelly touched the dial on the Channel Master control box, which "was marked like a compass," Gomer faded from view until Mama "carefully turned it back to the south. Gomer reappeared, resurrected," and the children were given firm orders: "We were not to touch the TV" or the antenna box. "And if we so much as looked at the knobs that controlled the color, she would whip us. It had taken her all afternoon to get the color just right."[37]

In later years the narrator comes to see television as a medium of illusion and groundless optimism; he hears one character on the sitcom "*Alice*" tell another, "'Buck up, kiddo, everything's going to be all right.' And what I'm trying to tell you now is this: I grew up in a split-level ranch house outside a town that could have been anywhere. I grew up in front of a television. I would have believed her."[38]

Novels and films have probed the degrees to which some especially susceptible viewers "become" what they see on television. Television is not a frequent subject of poets, but scattered poems link TV with states of being too. Shel Silverstein's lighthearted "Jimmy Jet and His TV Set" pictures an avid viewer who metamorphoses into a TV set,[39] while Elizabeth Bishop's somber poem "House Guest" depicts a foreign visitor who stares hopelessly at the zigzag patterns on an unadjusted TV.[40] In his paean "To Television," Robert Pinsky, who served a term as Poet Laureate of the United States, describes the set housed in an armoire in his hotel room, ready to provide distractions of entertainment and memory shortly before he goes to face his own audience.[41]

TV in Autobiographies and Essays

Library and bookstore shelves offer a variety of autobiographies of women and men professionally involved in television, but the most interesting insights about TV often occur in the writings of those who live outside the broadcast industry. Lucy Grealy came to public attention in reports of facial disfigurements caused by a rare cancer and the medical operations meant to treat it, and, to protest unwanted media attention, she ruefully titled a book of essays, *As Seen on TV*. Reference to television programs and heroes is a unifying thread in David Benjamin's *The Life and Times of the Last Kid Picked*, and recollections of childhood viewing link stages of Amy Blackmarr's life in *Going to Ground: Simple Life on a Georgia Pond*.

According to the opening of the essay "Folding the Times," George W. S. Trow's father taught him the "almost useless" skill of folding a large-format newspaper so that it could be read in a cramped subway seat,[42] but the discussion moves toward a distinction between the ways that print and broadcast news sources have reshaped the public's sense of "news." "Television will not allow you to follow a story," Trow says; "Each broadcast is self-contained; the television news people are embarrassed if they ever have to remind you that the story existed yesterday as well. The only exception is Big Human Interest. If it has the quality of a soap opera—O. J. Simpson, or the plane that exploded mysteriously—then they trust it as a story that will have had the dramatic elements necessary for their formula." Thus television, the "ignorant little snipped of a medium" just emerging in 1950, has redefined newsworthiness: "There is no story on the front page of the *Times* of February 1, 1950 . . . that is related in any way, shape, or form to the world that is being reported on today by the panels of communications experts you can watch on C-SPAN or CNN." Trow's contemplation of "our current culture of irony and anger and freneticism—our TV culture"[43] leads to a stern judgment: "My view of the civilization as it was presented in the *Times* of February 1950 is that in the Second World War the Germans lost and television won."[44]

In 1974 Donald Barthelme, author of wacky postmodern novels and short stories, published *Guilty Pleasures*, a collection of essays and parodies including "And Now Let's Hear It for the Ed Sullivan Show!" At first the essayist seems merely to be logging each shot and cutaway of a typical evening on the era's most-watched variety program: "The Ed Sullivan Show. Sunday night. Church of the unchurched. Ed stands there. He looks great. . . . Sways a little from side to side."[45] Choppy utterances simulate the restless switching of camera angles and simultaneously suggest the speaker's claustrophobically close focus on the screen, as if his face is only inches away and he dares not blink for fear of missing something. The essay's style conveys the notion that such a TV hour is at once too much and too little: the excitements of the changing acts and the audience responses urged by the host ultimately fall into a sameness, as each cutaway demands a new focus of attention. Television, the essay implies, equalizes everything into snippets finally amounting to a slightly surreal flatness.

W. D. Wetherell begins his essay "On the Road and on the Tube" by declaring himself a reluctant viewer except in one kind of setting. "We haven't had a television set in our house since 1983," he says, "when my wife and I took our small, rabbit-eared black-and-white to the local dump and heaved it toward a mound of old paint cans and broken screens," but motel rooms are another matter: "At home, you *watch* TV; in motels, you *wallow* in it, let it knock you over the head until you're sleepy." This occasional viewing "is part of a cultural refresher course that usually ends up leaving me ... refreshingly appalled. *My word*, I feel like saying. *This* goes on now!" Wetherell laments that for a time local TV was lost among the cable and satellite program choices, but community access channels and C-SPAN roundtables have brought some sense of real community back to the motel TV screen.[46] This is not a medium that he wants to live with, but it is a guilty pleasure to be absorbed among the expected displacements of traveling.

Notes

1. "Pioneers: William Golden," *Communication Arts 287* (March–April 1999): 180–181.

2. Cipa Pineles Golden, Kurt Weihs, and Robert Strunsky, *The Visual Craft of William Golden* (New York: George Brazillier, 1962), 64–65.

3. "Paul Strand at 82," *Photography Year 1973* (New York: Time-Life Books, 1972), 13.

4. CBS Television advertisement, *Fortune* 40 (September 1949): 58–59.

5. "The Saturday Evening Post," in Maureen Hart Hennessey and Anne Knutson, eds., *Norman Rockwell: Pictures for the American People* (New York: Harry Abrams, 1999), 149.

6. Bruce Kurtz, *Portfolio* (May–June 1982): 100–103.

7. Ibid., 101.

8. Lloyd DeGrane, *Tuned In: Television in American Life* (Urbana and London: University of Chicago Press, 1991), n.p.

9. Ibid.

10. Mark Bennett, *TV Sets: Fantasy Blueprints of Classic TV Homes* (New York: TV Books), 1996.

11. Grady T. Turner, "Mark Bennett at the Corcoran," *Art in America*, 86(2) (February 1998): 110.

12. Reprinted in Joseph J. Corn and Brian Horrigan, *Yesterday's Tomorrows: Past Visions of the American Future*, ed. Katherine Chambers (New York: Summit Books and Washington, DC: Smithsonian Institution Traveling Exhibition Service, 1984), 25.

13. http://www.familycircus.com/mylife.html, accessed June 14, 2004.

14. "Obituaries for TV Show Characters," *Mad* (194); quoted from Maria Reidelbach, *Completely Mad: A History of the Comic Book and Magazine* (Boston, MA: Little, Brown, 1991), 76.

15. Reidelbach, *Completely Mad*, 72.

16. Robert Crumb, *The R. Crumb Coffee Table Art Book*. Paperback edition. (N.p.: Back Bay/Kitchen Sink, 1998), 26.

17. Corn and Horrigan, *Yesterday's Tomorrows*, 24.

18. Ibid., xv.

19. Ibid., 25.

20. Quoted in "Writer Shapes 'Paranoia' into Film," *Greenville News* (June 5, 1998): "Time Out" tab sec.

21. "Cinema: Shading the Past," *Time* (October 26, 1998): 92.

22. "Observer: Saps of Now and Then, *New York Times* (November 27, 1998): sec. A.

23. The speech is transcribed here from the soundtrack of the MGM/UA Contemporary Classics VHS release. It should be noted that minor changes of this speech's wording occur in Sam Hedrin's paperback novelization of Chayefsky's script, *Network* (New York: Pocket Books, 1976), 142.

24. The last sentence in Hedrin's novelization is "Television continued relentlessly on." Hedrin, 149.

25. http://www.turnerclassicmovies.com/ThisMonth/Article/0,1185.0.html, accessed June 9, 2004.

26. October 29, 2000, http://www.suntimes.com/ebert/greatmovies/network.html.

27. Wally Lamb, *She's Come Undone* (New York: Pocket Books, 1993), 3.

28. Ibid., 5.

29. Curtis White, *Memories of My Father Watching TV* (Normal, IL: Dalkey Archive Press/Illinois State University, 1998), 3.

30. Ibid., 74.

31. Ibid., 4.

32. Marjorie Klein, *Test Pattern* (New York: Perennial, 2001), n. p.

33. Ibid., 6–7.

34. John Grisham, *A Painted House* (New York: Doubleday, 2000), 269–270.

35. Ibid., 272.

36. Tony Early, *Somehow Form a Family* (Chapel Hill, NC: Algonquin Books, 2001), 1.

37. *Ibid.,* 6–7.

38. Ibid., 18.

39. Shel Silverstein, *Where the Sidewalk Ends: The Poems and Drawings of Shel Silverstein* (New York: Harper and Row, 1974), 28–29.

40. Elizabeth Bishop "House Guest," in *The Complete Poems* (New York: Farrar, Straus and Giroux, 1969), 202.

41. Robert Pinsky, *Jersey Rain* (New York: Farrar, Straus and Giroux, 2000), 31–32.

42. George W. S. Trow, "Folding the Times," *The New Yorker* (December 28, 1998–January 4, 1999): 48.

43. Ibid., 52.

44. Ibid., 55–56.

45. Donald Bartholme, *Guilty Pleasures* (New York: Farrar, Straus and Giroux, 1974), 101.

46. W. D. Wetherell, "On the Road and on the Tube" *New York Times* (September 7, 1997): sec. C.

Bibliography

Anderson, Tim. "Go Play." *Herald-Leader* (Fitzgerald, GA) (October 4, 2006): sec. B.

Baker, Russell. "Observer: Saps of Now and Then." *New York Times* (November 27, 1998): sec. A.

Baldwin, James. *Nobody Knows My Name: More Notes of a Native Son.* New York: Dial Press, 1961.

Barfield, Ray. *Listening to Radio, 1920–1950.* Westport, CT, and London: Praeger, 1996.

Barlow, Perry. Untitled Cartoon. *Collier's* 134(1) (July 9, 1954): 84.

Barron, James. "On One TV Channel, Impostors Get Respect." *New York Times* (January 16, 2002): sec. B.

Barthelme, Donald. *Guilty Pleasures.* New York: Farrar, Straus and Giroux, 1974.

Benjamin, David. *The Life and Times of the Last Kid Picked.* New York: Random House, 2003.

Bennett, Mark. *TV Sets: Fantasy Blueprints of Classic TV Homes.* New York: TV Books, 1996.

Bishop, Elizabeth. *The Complete Poems.* New York: Farrar, Straus and Giroux, 1969.

Blackmarr, Amy. *Going to Ground: Simple Life on a Georgia Pond.* New York: Penguin, 1998.

Brinkley, Joel. "Defining TV's and Computers for a Future of High Definition." *New York Times* (December 2, 1996): national edition, sec. C.

Brooks, Jeanne. "See Some Good Stories Saturday." *Greenville (South Carolina) News* (August 14, 2003): sec. B.

Bruni, Frank. "Sheldon Leonard, 89, Actor and Producer of TV Hits, Dies." *New York Times* (January 13, 1997): sec. A.

Bryson, Bill. *The Life and Times of the Thunderbolt Kid: A Memoir.* New York: Broadway Books, 2006.

"But Do We Like What We Watch?" *Life* 71(11) (September 10, 1971): 40–44.

Butler, Sherri. "Accidental Learning." *Herald-Leader* (July 11, 2001): sec. B.

Calvo, Dana. "'The President's Been Shot.'" *Smithsonian* 34(8) (November 2003): 80–88.

Carpenter, Allan. "Where Is Television?" *Science Digest* 20(6) (December 1946): 54–57.

Collins, Philip. *The Golden Age of Televisions.* Santa Monica and Los Angeles, CA: General Publishing Group, 1997.

Collins, Ronald K. L. "TV Subverts the First Amendment." *New York Times* (September 19, 1987): 18.

Columbia Broadcasting System advertisement "Harvest." *Fortune* 40(2) (September 1949): 58–59. Reprinted in *Communication Arts 287* (March-April 1999): 181.

Corcoran, Penelope. "Encounter with 'The French Chef' Made a Lasting Impression," *Seattle Post-Intelligencer* August 14, 2004, http://seattlepi.nwsource.com/food/186266_juliaconnection14.html.

Corliss, Richard. "Cinema: Shading the Past." *Time* 152(17) (October 26, 1998): 92.

Corn, Joseph J., and Brian Horrigan. *Yesterday's Tomorrows: Past Visions of the American Future.* Edited by Katherine Chambers. New York: Summit Books; Washington, DC: Smithsonian Institution Traveling Exhibition Service, 1984.

Cronkite, Walter. Interview Segment. "Television: Window to the World," *Modern Marvels.* The History Channel, first broadcast 1999; repeated September 3, 2003.

Crumb, Robert. *The R. Crumb Coffee Table Art Book.* N.P.: Back Bay/Kitchen Sink, 1998.

Current Biography Yearbook 1986. New York: H.W. Wilson Co., 1987.

DeGrane, Lloyd. *Tuned In: Television in American Life.* Urbana and Chicago: University of Illinois Press, 1991.

Doherty, Thomas. "Assassination and Funeral of President John F. Kennedy." The Museum of Broadcast Communications,. http:www.museum.tv/archives/etv/K/htmlK/Kennedyjf.htm.

Earley, Tony. *Somehow Form a Family: Stories That Are Mostly True.* Chapel Hill, NC: Algonquin Books, 2001.

Ebert, Roger. Review of *Network. Chicago Sun-Times,* http://www.suntimes.com/ebert/great movies/network.html.

———. Review of *Quiz Show. Chicago Sun-Times* (September 16, 1994), http://www.suntimes.com/ebert_reviews/1994/09/940795.html.

Eliot, T. S. Letter to the Editor. *Times* (London) (December 20, 1950): 7.

Fant, Reese. "If You've Got Any Radio Stories, There's Someone Who Wants Them," *Greenville News* (September 19, 1993): sec. D.

Farrow, David. "Elephant Antic Was Set Up, Reader Says." *Post and Courier* (Charleston, SC) (October 24, 2002): sec. D.

———. "Happy Raine Shares Her Memories of Early Days of TV in Charleston." *Post and Courier* (October 10, 2002): sec. A.

———. "Local TV Newsman Airs Polka Dots on Broadcast." *Post and Courier* (November 14, 2002): sec. F.

———. "Reader Recalls Early Days at Channel 5 with Charlie Hall." *Post and Courier* (November 7, 2002): sec. F.

———. "Reader Recalls Short Story Has Suzie-Q Sent to Atlanta Zoo." *Post and Courier* (October 3, 2002): sec. F.

———. "Remember When? Children Delighted in Suzie-Q." *Post and Courier* (August 29, 2002): sec. F.

———. "Story Involves a Governor, an Elephant, the Press and Flies." *Post and Courier* (July 3, 2003): sec. D.

———. "Truth Told about Newscaster in Boxers." *Post and Courier* (November 21, 2002): sec. F.

Feagler, Dick. "Blow to Advertising Revenue May Restore Responsibility to TV." *Greenville News* (November 17, 1995): sec. A.

Fields, Mitch. "Incoming: Nowhere to Surf." *New York Times* (February 18, 1999): Circuits sec.

Fisher, David E. and Marshall Jon. *Tube: The Invention of Television.* Washington, DC: Continuum, 1996.

Frankel, Max. "Word & Image: Pay-Pay-Pay-Per-View." *New York Times Magazine* (November 4, 1998): 36.

"Fred W. Friendly, CBS Executive and Pioneer in TV News Coverage, Dies at 82." *New York Times* (March 5, 1998): sec. B.

Galbraith, John Kenneth. *The New Industrial State.* Boston, MA: Houghton Mifflin, 1967.

Gitlin, Todd. *The Twilight of Common Dreams: Why America Is Wracked by Culture Wars.* New York: Metropolitan Books/Henry Holt, 1995.

Glasheen, Leah K. "Reconsidering the '50s." *AARP Bulletin* 39(1) (January 1998): 20, 8.

Golden, Cipe Pineles, Kurt Weihs, and Robert Strunsky, eds. *The Visual Craft of William Golden.* New York: George Braziller, 1962.

Grisham, John. *A Painted House.* New York: Doubleday, 2000.

Hainer, Cathy. "Turning Off the TV Gets Woman Plugged into Living." *Greenville News* (July 4, 1997): sec. B.

Harris, Jay, with *TV Guide* editors. *TV Guide: The First 25 Years.* New York: Simon and Schuster, 1978.

Hedrin, Sam. *Network.* New York: Pocket Books, 1976.

Heimann, Jim, ed. *All-American Ads of the '60s.* Koln, Germany: Taschen, 2000.

Hennessey, Maureen Hart, and Anne Knutson, eds. *Norman Rockwell: Pictures for the American People.* New York: Harry Abrams, 1999.

Hill, Ric and Stebbo. *TV Photographer, 1957–1962.* Macon, GA: An Obscure Publication; Atlanta: Southeastern Arts Media and Education Project, 1988.

Hine, Thomas. "Commercials in Flux: What's around the Corner for Ring around the Collar?" *New York Times* (May 30, 2004): sec. D.

Johnson, Rheta Grimsley. "Cajuns Are the Most Hospitable of Southerners." *Atlanta Journal-Constitution* (December 2, 1998): sec. C.

———. "Daytime TV Won't Cure Flu, But It Helps." *Atlanta Journal-Constitution* (February 7, 2000): sec. E.

———. "Pizza on Roof Could Use Some Pepperoni." *Atlanta Journal-Constitution* (September 18, 1998): sec. C.

Kaplan, Jerry. *Startup: A Silicon Valley Adventure.* Boston, MA: Houghton Mifflin, 1995.

Kirschner, Ann. Letter to the Editor. *New York Times* (August 17, 2004): sec. A.

Kisseloff, Jeff. *The Box: An Oral History of Television, 1920–1961.* New York: Viking, 1995.

Klein, Marjorie. *Test Pattern.* New York: Perennial, 2001.

Kloer, Phil. "Channel Surfer: Murphy Often Told Us What We Needed to Hear." *Atlanta Journal-Constitution* (May 18, 1998): sec. D.

Krebs, Albin, and Robert McG. Thomas, Jr. "Notes on People: TV Turns Off Its Father." *New York Times* (July 31, 1981): final city edition, sec. B.

Kurtz, Bruce. "The Zen Master of Video." *Portfolio* 4(3) (May–June 1982): 100–103.

Lamb, Wally. *She's Come Undone*. New York: Pocket Books, 1993.

Lippert, Barbara. "The Image: About Last Nite." *New York* 30(15) (April 21, 1997): 20–21.

Lynch, Lorrie. "Who's News." *USA Weekend* (July 31–August 2, 1998): 2.

McCann, Dennis. "Nation Saw McCarthy at His Worst on Murrow's CBS Broadcast," *Milwaukee Journal-Sentinel* (October 1, 1998), http://www.jsonline.com/news/wis150/stories/101sesq.stm.

McCrohan, Donna. *Prime Time, Our Time: America's Life and Times Through the Prism of Television*. Rocklin, CA: Prima Publishing & Communications, 1990.

McGeehan, Bernice. "Lawrence Welk: Champagne and Grace Notes." *Saturday Evening Post* 252(2) (March 1980): 52–56, 84–87.

McKee, Myrna. "Meeting Mister Television." *Daily Journal* (Seneca, SC) (December 14, 2000): sec. A.

McNeil, Alex. *Total Television: The Comprehensive Guide to Programming from 1948 to the Present*. 4th ed. New York: Penguin, 1996.

Miller, Jeff. "A U.S. Television Chronology, 1875–1970," http://members.aol.com/jeff560/chronotv.html (accessed August 12, 2003).

Minow, Newton N. "Excerpts from Speech by Minow." *New York Times* (May 10, 1961): sec. A.

Minow, Newton N, John Bartlow Martin, and Lee M. Mitchell. *Presidential Television: A Twentieth Century Fund Report*. New York: Basic Books, 1973.

"Mummified Body of German Found in Front of TV Years Later." *Greenville News* (June 27, 1998): sec. A.

Murrow, Edward R. "Views and Reviews: A Broadcaster Talks to His Colleagues." *The Reporter* 19(8) (November 13, 1958): 32–36.

Nemy, Enid, ed. "Metropolitan Diary." *New York Times* (February 7, 1999): 45.

Newcomb, Horace, ed. *Museum of Broadcast Communications Encyclopedia of Television*. 3 vols. Chicago and London: Fitzroy Dearborn, 1997.

Nixon, Rob. "Turner Classic Movies This Month," http://www.turnerclassicmovies.com/ThisMonth/Article/O,1185,0.html.

Noble, Elaine. "When 'Window Movies' Came to Des Moines." *Reminisce* 13(4) (July/August 2003): 12–13.

Noble, Holcomb B. "Frank Stanton, Who Guided CBS into the Television Era, Dies at 98." *New York Times* (December 26, 2006): sec. A, C.

"Oxevene Richardson Rites Held." *Herald-Leader* (April 11, 2007): sec. A.

Oxford, Edward. "TV's Wonder Years." *American History* 30(6) (January–February 1996): 26–31, 68–69.

Pinsky, Robert. *Jersey Rain*. New York: Farrar, Straus and Giroux, 2000.

Pitts, Leonard. "Humans Can't Bear Much Reality." *Herald-Leader* (June 30, 2004): sec. B.

———. "TV Keeps Us on the Same Wavelength." *Atlanta Journal-Constitution* (April 30, 1998): sec. A.

Powell, Kay. "Leona McWilliams, 105, Widowed Farmer." *Atlanta Journal-Constitution* (October 12, 2001): sec. D.

Reidelbach, Maria. *Completely Mad: A History of the Comic Book and Magazine*. Boston, MA: Little, Brown, 1991.

Rogers, Joe, ed. "Metropolitan Diary." *New York Times* (August 25, 2003): sec. A.

———. "Metropolitan Diary." *New York Times* (May 24, 2004): sec. A.

Schenck, Ernie. "Advertising: A Case of Winky Dink Envy." *Communication Arts 329* (July 2004): 184.

Seplow, Stephen, and Jonathan Storm. "The Little Box at the Core of Our Culture." *Atlanta Journal-Constitution* (January 11, 1998): sec. H.

Sibley, Celestine. "A Class Act Should Decide Person's Status." *Atlanta Journal-Constitution* (April 5, 1999): sec. C.

Silverstein, Shel. *Where the Sidewalk Ends: The Poems and Drawings of Shel Silverstein.* New York: Harper and Row, 1974.

Skube, Michael. "If Television's Own Executives Trash It, Who Are We to Disagree?" *Atlanta Journal-Constitution* (September 16, 1997): sec. B.

Smith, Martha. "Magic Screen of Television Boon to Family Life in Atlanta." *Atlanta Journal-Constitution* (September 23, 1951): sec. F.

Smith, Nelson. "Lives: Secret Admirers." *New York Times Magazine* (August 23, 1998): 64.

Stanley, Alessandra. "Class Issues: TV's Busby Berkeley Moment." *New York Times* (January 30, 2005): sec. D.

Teachout, Terry. "The Games People Played in a Simpler Time." *New York Times* (October 28, 2001): sec.D.

"Television." *NARA News* 25(3) (Summer 1997): 33.

"Television: Window to the World," *Modern Marvels.* The History Channel, 1999; repeated September 3, 2003.

"Television's Big Brother" (CBS Radio advertisement). *New York Times* (June 18, 1951): sec. C.

Tichi, Cecelia. *Electronic Hearth: Creating an American Television Culture.* New York and Oxford: Oxford University Press, 1991.

Time-Life Books editors. *Photography Year 1973.* New York: Time-Life Books, 1972.

Trow, George W. S. "Folding the Times." *The New Yorker* 84(40) (December 28, 1990; January 4, 1999): 48–57.

Turner, Grady T. "Mark Bennett at the Corcoran." *Art in America* 86(2) (February 1998): 110.

"Update: Walter Cronkite Still at the Helm." *Newsweek* 102(23) (December 5, 1983): 24–28.

Van Buren, Abigail. "Dear Abby." *Atlanta Journal-Constitution* (February 7, 1997): sec. H.

"The Vent." *Atlanta Journal-Constitution* (January 1, 1997): sec. A.

"The Vent." *Atlanta Journal-Constitution* (May 7, 1999): sec. D.

Wakin, Edward. *How TV Changed America's Mind.* New York: Lothrop, Lee & Shepard Books, 1996.

Wallis, Claudia. "The Artificial Heart: Act III." *Time* 125(9) (March 4, 1985): 69.

Wetherell, W. D. "On the Road and On the Tube." *New York Times* (September 7, 1997): sec.E.

White, Curtis. *Memories of My Father Watching TV.* Normal, IL: Dalkey Archive Press/Illinois State University, 1998.

White, E. B. "One Man's Meat." *Harper's* 177 (October 1938): 553–556.

[White, E. B.]. "Talk of the Town: Television." *The New Yorker* (December 4, 1948): 25–29.
———. *Writings from The New Yorker, 1927–1976.* Edited by Rebecca M. Dale. New York: HarperCollins, 1990.

Winn, Marie. "The Plug-In Drug." *Saturday Evening Post* 249 (November 1977): 40–41, 90–91.

———. *The Plug-In Drug.* New York: Viking, 1977.

Winship, Michael. *Television.* New York: Random House, 1988.

"Writer Shapes 'Paranoia' into Film." *Greenville News* (June 5, 1998): Time Out sec.

Yacowar, M. Review of David Marc, *Bonfire of the Humanities: Television, Subliteracy, and Long-Term Memory Loss. Choice* 33(3) (November 1955): 456.

Films

Network. DVD. Warner Home Video, 2000.

Pleasantville. VHS. New Line Productions, 1999.

Quiz Show. VHS. Hollywood Pictures Home Video, 1995.

The Truman Show. VHS. Paramount Pictures Home Video, 1999.

Interview

Lotus, Jean. Telephone interview, June 9, 2004.

Web Sites

http://ktla.trb.com

http://members.fortunecity.com/drg45nzp/hallictv.html

http://www.familycircus.com/mylife/mylife.html

http.www.richsamuels.com/nbcmm/wmaq/w9xap/w9xapnewsreel.html

Index

About the Author

RAY BARFIELD is Professor of English at Clemson University and is the author of *Listening to Radio, 1920–1950* (Praeger 1996). He is a contributor to *The Greenwood Guide to American Popular Culture* and *The Museum of Broadcasting Encyclopedia of Radio*, and he has authored numerous articles and conference papers on radio, cartoons, children's books, and 1920s fads. He is also coauthor of a business communications handbook.